ROMANCING THE VINE

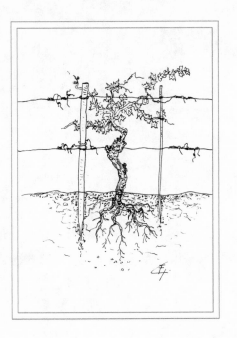

ALAN TARDI

ROMANCING

THE

VINE

LIFE, LOVE &

TRANSFORMATION

IN THE VINEYARDS

OF

BAROLO

NEW YORK

St. Martin's Press

www.stmartins. com

Book design by Fritz Metsch
Illustrations by Pierflavio Gallina
Maps by David Cain

Library of Congress Cataloging-in-Publication Data

Tardi, Alan, 1956–
    Romancing the vine : life, love, and transformation in the vineyards of Barolo / Alan Tardi.—1st ed.
        p.   cm.
    ISBN-13: 978-0-312-35794-8
    ISBN-10: 0-312-35794-X
    1. Castiglione Falletto (Italy)—Description and travel.   2. Castiglione Falletto (Italy)—Social life and customs.   3. Tardi, Alan, 1956—Travel—Italy—Castiglione Falletto.   4. Cookery, Italian.   5. Wine and wine making—Italy—Castiglione Falletto. I. Title.
DG975.C2695625T37 2006
945'.13—dc22   2006008202

FIRST EDITION: NOVEMBER 2006

10   9   8   7   6   5   4   3   2   1

ACKNOWLEDGMENTS

I WOULD LIKE to express gratitude to my friend and agent, David Black, for his unfailing support and encouragement, and to my editor, Elizabeth Beier, and the staff at St. Martin's for their thoughtful advice and expertise in shepherding the manuscript to publication. Special thanks to editorial assistant Michelle Richter, production editor Robert Cloud, and Karen Gillis, director of production. Thanks also to my sister, Carla Tardi, for her insightful reading of a pre-publication draft.

Thank you, Bruna, Renza, Gemma, Domenica, and Rita; Bruno at Ristorante Moderno, Cristina at Ristorante Le Torri, Signora Costa at Café Bicerin, and to everyone else who generously supplied recipes, culinary wisdom, and lots of good food. Thanks also to Siobhan Thomas and Jill Sloane, who tested the recipes.

Thanks to my friend and neighbor Pierflavio Gallina for the wonderful illustrations, Pasquale Catanzariti for taking my picture in the vineyards, and Bruno Murialdo for the jacket photo.

Thanks to all the winemakers and farmers of Barolo who shared their viticultural knowledge and love for the land, especially Marco, Tiziana, Francesco, Enrico, and Giovanni at Cantina Parusso and Alfredo, Luciana, and Luca at Cantina Vietti, and to all my neighbors in Castiglione Falletto, who tolerated a stranger in their midst and gradually admitted me into their world.

A special thanks to Silvana and Renata of Hotel Le Torri for their "eccentric" hospitality, to Patricia at the *municipio* for processing my citizenship application, and to the Alessandria family of the Bussia.

Thanks to Fabrizio for his friendship and for letting me help in Le Munie and to Tiziana Mascarello for her acceptance and hospitality.

And, finally, heartfelt thanks to Ivana for the most precious gift of all.

CONTENTS

PART III: SUMMER INTERLUDE

PART IV: AUTUMN BOUNTY

PART V: DORMANCY

PART VI: FLOWERING AGAIN

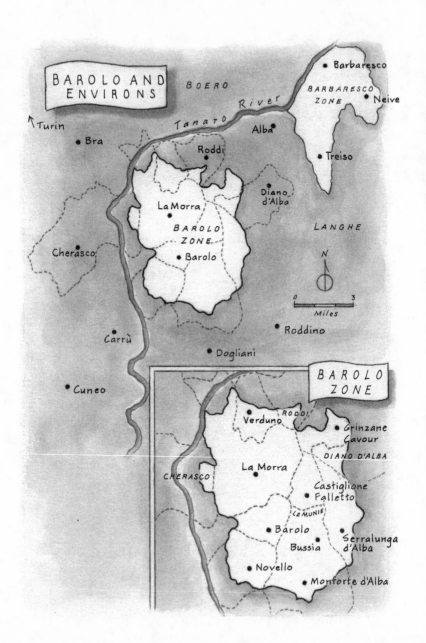

BAROLO AND ENVIRONS

BOERO

↑Turin

Tanaro River

Bra

Roddi

Alba

Barbaresco

BARBARESCO ZONE

Neive

Diano d'Alba

Treiso

La Morra

BAROLO ZONE

LANGHE

N

Cherasco

Barolo

0 — 3
Miles

Roddino

Carrù

Dogliani

Cuneo

BAROLO ZONE

Verduno

RODDI

Grinzane Cavour

DIANO D'ALBA

CHERASCO

La Morra

Castiglione Falletto

LE MUNIE

Barolo

Serralunga d'Alba

Bussia

Novello

Monforte d'Alba

ROMANCING THE VINE

*The heart's castle hangs high in the heavens, shrouded in
scudding clouds and obscured by mist.*

—THOMAS LEWIS, FARI AMINI, AND RICHARD LANNON
A General Theory of Love

Castle in the Clouds

CASTIGLIONE FALLETTO IS a tiny village in Italy, perched on a hill
cradled by vines, right smack in the heart of Barolo. The name—it
means "Big Castle of the Falletti"—is apt: the ancient square fortress sits
large and squat above the other small buildings that sprang up around it,
much as the village itself rests in obstinate stoniness atop the surrounding vineyards. As you drive out of the bustling mini-metropolis of Alba
in the direction of Monforte, the castle appears hovering on the horizon,
its stocky central tower rising up like a headless phallus through the mist.

The Falletti were a noble family who prospered here hundreds of
years ago. They carved vineyards and farmland out of virgin forests,
they constructed churches, roads, and castles, and around those castles
(and the peasant houses that clustered about them like metal filings
around a magnet), they built fortifications to protect themselves and
their subjects from invaders. According to local legend, the Falletti also
helped create a network of subterranean tunnels connecting neighboring castles and *palazzi* in case they needed to escape a siege or give asy-
lum to fellow nobles.

Today the Falletti are gone, but the big castle remains. It is privately
owned, supposedly by an elderly countess, though I have yet to see any
sign of life there except for once a year when it's opened up for tours.

Castiglione is a sleepy little village. It has a few hundred inhabitants,
a post office, and a bank (which is usually closed), a hotel, a restaurant, a

trattoria, a tobacco shop, a *panetteria,* and two bars. One of the bars, located at the foot of the village just off the Alba-Monforte road, opens early in the morning attracting a steady stream of passing workers who jump down out of their tractors and trucks for a quick espresso. The other, farther up in the middle of the village, opens later but stays open later too, and this is where the locals go in the evening to hang out and play cards. Most of the businesses are located on a little piazza at the center of town where old people gather in the afternoon to gossip and take the sun. On Sunday mornings, a respectable portion of the population makes the pilgrimage up the hill to the lovely little church (also built by the Falletti) next to the castle.

It's a beautiful, if unremarkable, place: there are fantastic views of the countryside, many of the narrow streets still have their old cobblestones, and the town has the peculiar resounding quiet characteristic of a medieval enclave. I suppose you could say the same about any number of villages in Piemonte—or in Italy or Europe, for that matter. But in the world of wine, tiny Castiglione is special; it has the distinction of being one of only three of the eleven communes of Barolo to be completely within the legal limits of the zone.[1] And for me, it has the further distinction of being the place I call home.

Unlike most of my neighbors, I wasn't born here. I'm not Piemontese; I'm not even officially Italian. I'm a chef (or at least I used to be), and it was food and wine that brought me here in the first place. For this is the land of ethereal white truffles and noble red wines, and I made my initial pilgrimage here many years ago to sample them at the source.

This unimposing little village got short shrift on my early visits. I know I came here, but the town itself is a blur. I dashed from one place to the next in a heady haze of red wine, raw meat, and countless bowls of buttery ravioli. It was the bigger, flashier places—Alba, Monforte, Barolo—that stood out, the whole trip sandwiched into a short break from my hectic big-city schedule.

I came back here just about a year ago. This time it was different, *very* different. Actually, Castiglione was probably pretty much as it always had been—it was I who had changed. Fifteen years had passed. And my life was kind of a mess.

Leveling the Field

I HAD A restaurant in New York.

It was a charming little place in a quaint three-story landmark building that stood out from all the other warehouse buildings on a nondescript side street in the up-and-coming Flatiron District. It had always been a restaurant. It started out in 1909 as a ritzy eatery catering to the well-heeled patrons of the original Madison Square Garden and the shoppers of Ladies' Mile. But over the years, the center of activity shifted uptown, the neighborhood changed, and the site deteriorated, first into a steak'n'chops joint, then a saloon, and finally a Russian nightclub and brothel.

By the time I found it, the place had been closed for a while and had the telltale odor—a combination of rancid vinegar, deep-fryer fat, and long-ago-exhaled smoke—that is the legacy of old, abandoned restaurants. It was dark and filthy. There were murky green walls, cracked vinyl banquettes, and vintage pool hall–style lights hung low over the dented mahogany bar with its thick Chicago railing. Even I had to admit the place was kind of creepy. But I could also feel the warmth, personality, and Old World charm cowering just beneath the superficial mess.

So I scraped together all my money and got some friends and family to invest. I cleaned the place up, got a friend to cover the walls in a gorgeous *stucco veneziano* that took him three weeks to apply by hand, and installed a beautifully tiled wood-burning oven in the dining room. I lit candles, played music, and sprayed air freshener to try to exorcise the ghosts. Then we opened the doors.

Right from the beginning, people responded to the warmth and sincerity. Many also said it felt just like a country restaurant in Italy, which struck me as kind of funny since it was basically just what had been there all along. I named the place Follonico after a hilltop village in Tuscany, where I had lived and worked some years before. In a review in *New York* magazine that came out three weeks after we opened, the restaurant critic Gael Greene suggested the name reminded her of "an act of love," which I took as a good omen (the review was a rave!).

Having my own business wasn't easy—far from it—but it was by all accounts a success. We got great press and, more important, developed an extremely loyal and appreciative clientele. Many people genuinely loved the restaurant. But over time it also took its toll. I was both chef *and* owner; when I wasn't in the kitchen cooking (which I usually was) I was in the office dealing with a host of other issues. It was never-ending and all-consuming, and after nearly ten years I was getting tired.

Kay, my companion in life and partner in work, had gotten fed up with the restaurant business and left several years before to pursue a career in floral and event design. Now I too had had enough. I loved the restaurant and was proud of what we had accomplished. But I was also exhausted, frustrated, sorely in need of a change. And, given the idiosyncratic nature of the beast I myself had created, I saw no way to do that other than to close.

It was a well-thought-out, calculated move. I leveraged a fifteen-year lease extension from my landlord and quietly found a prospective buyer. After hands were shaken all around, I chose a date, had a tearful meeting with the staff, promising to try to find them jobs wherever they wanted, and sent out a mailing to our customers and a notice to the press. It was sad in some ways, but mostly I felt positive, excited, and relieved. On Bastille Day 2001 we served our last meal. Afterward, we had a big party for friends and staff. Then I closed the doors for good.

Kay and I set off on an extended trip to the beach. It seemed as if a weight had been lifted from my shoulders; I felt a whole new sense of freedom and lightness and was almost giddy with excitement at the wealth of possibilities that lay ahead. I came back in September sun-tanned and refreshed and ready to wrap up the restaurant deal. Then something else happened, something that was decidedly *not* calculated—at least not on my part.

Kay and I were having coffee one morning, and I was sitting in my chair looking out the window.

My apartment is on a high floor in a prewar Chelsea building with great light and a wall of southwest-facing windows that looks down on the tip of lower Manhattan. On a clear day you can see all the way down to the Statue of Liberty. And this day was clear indeed, a perfect end-of-summer morning. The sky was bright blue; though still early, the sun was already out in full force and from the total lack of humidity in the air you could tell that fall was just around the corner. It was uncharacteristically peaceful for New York, almost tranquil.

Then all hell broke loose. As I sat there, I noticed smoke coming from one of the towers of the World Trade Center. "What is that?" I said as I jumped up to go look through the telescope. At first it looked like the incinerator exhaust that buildings emit from time to time but thicker, and I had never seen it coming from there before. As I zeroed in, I noticed that not only was smoke now billowing out of the top, there was also what appeared to be a long horizontal hole in the building itself about three quarters of the way up. We turned on the television to try to find out what was going on. I was looking through the telescope when the second plane hit. . . .

We didn't know how yet—I'm not sure we do even now—but right away we knew that something monumental had happened and that the city, the world, our lives, would never be the same.

The following spring Kay and I decided to break up. It was as civilized as it could possibly be—no screaming or name calling, no affairs—but that didn't make it any easier.

Summer was coming up fast. For the past decade, each year we had scurried at the last minute to find a vacation house to rent for the weeks the restaurant was closed. This time, ironically, we had planned well ahead, even going out to the Cape the previous winter to look at houses. The two-bedroom Truro cottage we had selected was rented for four weeks and fully paid for in advance. I was convinced we could continue our relationship discussions while on vacation. But Kay disagreed, saying she really needed a break and some distance. So I stocked up on books, bought an Intermediate French course on tape, rented a digital piano, and set out alone.

When I returned to New York—on September 11, 2002, of all days—Kay was gone. She had told me the week before that she felt she needed to have a place of her own and had taken an apartment uptown. I knew she was not going to be at my place. But all the same, I was unprepared for the feeling of utter emptiness when I walked in the door of what had been our home.

No restaurant, no World Trade Center, no "significant other"—it seemed as if my world too had collapsed.

I was supposed to go back to Italy in late October to write a magazine article about a food exposition in Torino. And after that I had agreed to go down to a villa in Tuscany to help some chef friends teach a cooking class. Now I really didn't want to go, didn't want to venture out of the protective cocoon I had spun around myself. But I also had a hard-to-come-by magazine assignment that I didn't want to pass up. So I dragged myself through the motions of booking a flight, finding a hotel, packing, getting myself to the airport. And I went.

PART I

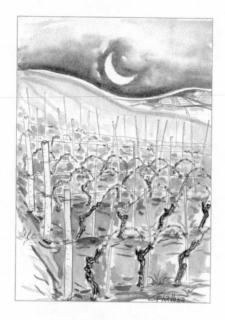

FIRST

PRUNING

WINTER

I · FIRST FORAYS

I HAD BEEN to Italy many times before, but this was my first visit to Torino. In many ways this city didn't feel like Italy at all. Wide, straight, portico-lined boulevards, square piazzas, and large traffic circles arranged with ordered symmetry into long vistas punctuated by prominent sculptures or imposing edifices seemed almost Parisian, while immense cafés of gleaming marble, sparkling crystal, and ornate stucco suggested Vienna.

The Salone del Gusto, a massive biennial celebration put on by a Piemonte-based organization called Slow Food, was different too.

Day after day, from 11 A.M. until well after dark, I walked up and down long corridors of stands set up by artisan food producers from all over Italy, Europe, and the world. I tasted sweet red garlic, salty prosciutto from a wild pig you have to slice as if playing a violin, and a dark, funky cheese that had been buried underground for a year. I went to workshops on a myriad of arcane subjects, such as a comparative tasting of four protected species of chicken (each simply boiled and served at room temperature without seasoning so as not to alter the natural flavor of the bird), another describing a traditional practice of raising rabbits in caves, and yet another featuring sheep's milk cheese made by Sardinian shepherds, paired with wines from the island's obscure indigenous grape varieties.

For a break, I strolled along the shopping arcade of the Renzo Piano–designed hotel next door and watched "normal" people devour hot dogs and slices of reheated pizza without giving it a second thought. On several occasions I took the elevator up to the mercifully small and uncrowded Agnelli Museum on the top floor and there gazed silently on the large, soberly precise paintings of Venice by

Canaletto, a series of early Matisses from the south of France—warm and busy and gay with bright though somewhat muted colors as if washed out by an abundance of sunlight—and my favorite: the cool, restrained but immensely sensuous *Nu, Couché* by Modigliani.

In the evenings when I could stand no more of the frenetic fair, I returned to the Centro Storico, the old historic part of town, and wandered through streets and plazas already decked out with a bedazzling array of holiday lights, each installation designed by a different artist.

After the fifth day I had more than enough information to write my story. My stomach was full to capacity with delicious unusual foods, my head was swimming with data, and my psyche was jittery. I was experiencing physical and emotional overload and needed to chill, preferably in a quiet place with someone familiar.

I was not expected in Tuscany for another ten days, so I decided to go visit a friend, Bruna, in the wine country near Alba. It was a mere hour to the southeast and I could stop off there for a day or two before continuing to work my way down to Carmignano.

Bruna is a great cook, especially when it comes to the regional dishes of her native Piemonte, and was one of the original "Mothers" of a restaurant in Manhattan called Le Madri, where I had been the original executive chef. I picked Bruna up at JFK when she first came to New York, helped her get acclimated to her new high-rise home and to working in a big Manhattan restaurant kitchen, took her to the dentist, and translated for her when hand gestures alone weren't sufficient.

We became friends and kept in touch after I left Le Madri. Some months later, Bruna left the restaurant too but stayed on in New York for nearly a decade, cooking in a number of different places. The year before, she had returned home to Piemonte for good and was now working in a restaurant in Castiglione, the town where she was born.

I picked up my rental car in Torino, got myself onto the autostrada, and, about an hour later, left the highway and zigzagged through the countryside to the little village. I met Bruna in the café at the foot of the hill. She brought me up to the hotel to see if they had any rooms available. It wasn't easy—it was still truffle season—but they managed to find me a room, as long as I didn't mind switching the next day.

I didn't mind at all. I was laid back and agenda-free (at least for the next ten days), just taking things as they came. Originally I had thought to stay only a couple nights but kept extending my visit incrementally, changing rooms not once but three times.

I liked Castiglione. It was cozy and quiet; the people were friendly, and, what's more, they seemed to like me too.

· RECIPE #1 ·

BICERIN

BICERIN (FROM THE Piemontese dialect diminutive of *bicer*, referring to the glass in which it is typically served) is a bittersweet concoction of rich chocolate, coffee, and cream. It was invented here in Torino during the heyday of the 1800s and remains the sweet beverage of choice.

This baroque city offers many grand cafés where bicerin may be consumed in elegant surroundings. But my favorite place is a sweet, cozy little tearoom where you are efficiently attended to by the prim proprietress and her staff of female assistants. It is called simply Bicerin and is located on the Piazza della Consolata, right across from the sanctuary of the same name.

Shortly after your order is conveyed to the kitchen, you will hear the sounds of a whisk scraping energetically against a metal bowl (they make each Bicerin to order the old-fashioned way), and before long a glass of dark steaming chocolate and frothy cream will be placed proudly before you. Nobody does it better, and, in my opinion, there is no finer place to enjoy a Bicerin than right here.

The recipe, according to the *padrona*, is ridiculously simple: "Take 1 cup of the very best hot chocolate you can find, mix it with 1 demitasse of the very best espresso (our private blend is 100 percent Arabica and ground for us especially) and scoop a hefty dollop of freshly whipped cream on top. Serve it in a glass. *Et voilà!*" (But if the recipe is so simple, why does the bicerin at Bicerin taste so much more delicious than anywhere else?)

· RECIPE #2 ·

FINANZIERA

Classic Innards Stew, adapted from Ristorante del Cambio, Torino

THIS RICH STEW is one of the classics of Piemontese *cucina borghese* (cooking of the urban upper classes), and Ristorante del Cambio in Torino, in continuous operation since 1754, is credited with having invented it.

While Cambio's baroque dining room may seem frozen in time, it shows no sign of fatigue: crystal and gold leaf sparkle in the candlelight, mirrors glimmer, linen is crisp, and flowers fresh and fragrant. Even the table where the first king of Italy, Vittorio Emanuele II, used to have lunch each day is still there and perfectly set as if waiting for him to appear.

Many today might turn up their noses at this old-fashioned dish. But, while it may be predominantly offal, it is not awful. Each of the exotic ingredients has its own distinctive shape and mild taste, which meld together into an altogether inoffensive creamy richness, a subtle symphony of muted flavors and textures. If you can accumulate enough of the ingredients to create a reasonable facsimile—or, better yet, ever find yourself in Torino and can secure a table at Cambio—I would encourage you to try it.

INGREDIENTS FOR FINANZIERA

2 tablespoons white wine vinegar

Coarse salt

1 calf's brain (about 3 ounces)

6 ounces veal sweetbreads

2 ounces spleen (bone marrow) from the spinal cord of a calf (see note below)

3 ounces cockscombs

6 ounces veal testicles

4 ounces butter

3½ ounces chicken breast, lightly pounded and cut into ½-inch pieces

3 ounces (1¼-inch slice) calf's liver, cleaned of any cartilage and cut into ½-inch pieces

4 scallions, sliced into thin
    rounds
2 ounces of small porcini
    mushrooms marinated in oil
    and vinegar
1 bay leaf
½ cup dry Marsala

1 teaspoon cider vinegar
1 tablespoon balsamic vinegar
1 tablespoon of veal demi-glace
    or concentrated chicken
    broth

FOR THE SEMOLINA

1 pint milk
1 tablespoon sugar
⅓ cup semolina flour
1 egg yolk

3 cups vegetable oil for frying
2 tablespoons flour
1 egg + 1 egg white, beaten
¼ cup bread crumbs

I. PREPARE THE FRIED SEMOLINA

1. Bring the milk to a boil. Add the sugar.

2. Gradually add the semolina flour and cook, stirring constantly, for about 10 minutes.

3. Remove from the heat and stir in the egg yolk.

4. Pour the mixture out onto a small greased cookie sheet or Teflon baking pan (the semolina should be about ½ to ¾ inch thick) and let cool. (May be made a day or two ahead to this point.)

5. Heat the oil in a high-walled, heavy-bottomed pot. Cut the semolina into diamonds about 1 inch long, dust with the flour, dip in the beaten egg, roll in the breadcrumbs, and fry in the hot oil.

6. When golden brown, remove them from the oil with a slotted spoon and blot on a paper towel.

II. PREPARE THE FINANZIERA

1. Bring a pot of water to boil. Add the white wine vinegar and a pinch of salt. Rinse the calf's brain thoroughly in cold water, add to the boiling water and simmer gently for 6 to 8 minutes. Remove with a slotted spoon and shock under cold water. Put the brain on a plate,

cover with plastic wrap, and refrigerate with a light weight (such as a small tray) on top. Repeat with the sweetbreads and spleen.

2. Cook the cockscombs in the simmering acidulated water until tender (about 15 minutes). Remove with a slotted spoon and rinse under cold water. Repeat with the testicles.

3. Cut the sweetbreads, spleen, and brain into 1½-inch pieces. Cut the testicles and cockscombs in half (if the cockscombs are small, leave them whole).

4. Melt half of the butter in a large nonstick sauté pan. On a medium-high heat, add the chicken breast and the liver, and cook until opaque on the outside (about 1 minute). Add the scallions and continue cooking, stirring occasionally, until they are soft and transparent (about 4 minutes).

5. Add the mushrooms and the bay leaf and all of the offal except for the brain, and sauté for about 5 minutes. Add the brain to the pan and toss.

6. Deglaze with Marsala. Add the cider and balsamic vinegar and the demi-glace or stock. Add the remaining chilled butter to thicken the sauce.

Remove the bay leaf, divide onto plates, and serve hot with the fried semolina.

YIELD: SERVES 4 TO 6 AS AN APPETIZER.

———————

NOTE: Some of the ingredients for this recipe—veal testicles, spleen, cockscombs, etc.—may be difficult to find. Ask a good butcher, who may be able to get them for you. Leave something out if you can't get it. Some people in Piemonte consider this a poor man's dish, a way of utilizing available ingredients, and make a Finanziera-like dish with all poultry meat and giblets, others add little meatballs or peas and a touch of cream.

2 · TWIN TOWERS

ALBERGO LE TORRI (The Towers) is a small, quirky, typically Italian operation. Though the name may have a familiar ring with ominous overtones for any New Yorker, in this case it refers innocently to the stone towers of the ancient castle right next door. There are three of them, and they are still standing more or less as they have for hundreds of years.

The hotel occupies a large old building in the center of Castiglione Falletto. It was originally a wealthy merchant's home (some say a monastery) that had been vacated and left to crumble before being purchased by a local construction firm, perfunctorily restored, and turned into a small hotel. It is innocuously furnished; simple but serviceable, comfortable but decidedly not luxurious and not too expensive either. What really sets Le Torri apart are its fantastic views, its excellent position right next to the campanile in the heart of this otherwise tranquil little village, and what was referred to in one write-up as its "eccentric" management.

The hotel is operated by two young women who resemble a real-life version of the Odd Couple.

Silvy is short and robust with thin brown hair that, when not tied up in a ponytail or barrette, falls just below her thick shoulders. Her speech is articulate and sometimes tinged with sarcasm or exasperation. She eats boiled vegetables and organic rice cakes, compulsively watches an American soap called *Beautiful* each afternoon and smokes light cigarettes whenever possible. On her nights off she attends a belly-dancing class in Torino, but at work her movements display the utmost control.

Silvy is intelligent, organized, and detail-oriented, and usually prefers dealing with the account books to the public. She would do just about anything for a friend, but puts up a stiff wall of formality with strangers and practically cringes each time the reception door opens and a new guest walks in. Silvy is *not* a people-person—at least not until you get to know her.

"Hmm. So. You would like a room?" she said when I first met her, barely deigning to look up from the reservation book. "Well, I do have one available, but it is not very nice," indicating a door just off the reception area. "It is dark and noisy—you hear all the sounds of people coming in and out—and you'll have to change in a day or so. But you can have it if you wish." I took it.

Renata is the exact opposite. She is tall, big-boned, and bleached blond. Her movements are as large and broad (one might even say clumsy) as she is good-natured. She is a collector and a consumer; she loves to accumulate things, and she does so naturally as if by gravitational pull. I went to her apartment in Alba once, and it was literally packed floor to ceiling with stuff, so packed you had to move a pile off the couch in order to sit down. Even the bathtub, I couldn't help noticing, was occupied. Her car is the same way; it contains books, tapes, boots, two umbrellas, underwear, a toothbrush, an empty wooden wine crate, and several cans of Coke—and that was *after* she cleaned it up!

Renata eats continuously (she is especially fond of hard salami, soft Gorgonzola, and pizza with onions) and especially enjoys eating with others. Renata loves to have people around her—friends or hotel guests, it really doesn't seem to matter—and people, in turn, are drawn to her like bees to a flower. At the most casual tourist inquiry, she will pull out maps and books of the region and create detailed (though impossible) itineraries of everything you absolutely must see; if she can, she will sweep you up and drag you all over the place herself. Or give her half a chance and she will flip through tall stacks of photos of her annual month-long trip to a mission in Kenya and gladly accept a small donation to the humanitarian cause. She is open and friendly and generous to a fault but helplessly disorganized: she often has to call herself to find out where she misplaced her cell phone and is as unable to clean up, keep track of accounts, or handle reservations as she is incapable of saying no.

In theory, these character differences of the two principals *could* be complementary and beneficial components of a successful partnership. But in reality, they often end up working at odds and try whenever possible to arrange their schedules so as not to be at Le Torri at the same time.

This small enterprise has only one employee, who provides something of a buffer and helps keep the business going. Ivana combines many of the best qualities of her two employers without their respective drawbacks: she is friendly and personable but meticulously organized as well; she can greet people with a smile when they come in *and* make sure they are charged correctly when they leave.

While Renata's lack of organization often tries her patience, Ivana generally prefers it to her other boss's despondency. In fact, the two of them are friends.

I heard about Ivana even before I met her. "Oh, you'll like her," Renata said breathlessly on the second day of my stay. "She knows a lot about wine. She's studying to be a sommelier."

Renata was right.

That night, the two of them dragged me out to a disco in Alba. It was Halloween (an "emerging" holiday in Italy), the place was completely packed, and I, not having had an opportunity to change after a day-long whirlwind tour with Renata, was sweating bullets in my heavy turtleneck.

After that, Ivana came to hang out at the hotel in the evenings with Renata and other friends who happened to drop by, creating an impromptu party—and sometimes I hung out too. I quickly became an honorary member of the gang, a sort of unofficial foreign dignitary (Renata liked to introduce me as the "*giornalista americano* who had a restaurant in front of the World Trade Center before it fell down"), and Ivana became my special diplomatic attaché, explaining the rapid-fire repartee and, when necessary, interpreting my comments for the group.

Before long I learned that Ivana was a childhood friend and neighbor of Silvana, who had also worked at Le Madri for a time. I mentioned it would be great to see her again, so Ivana offered to take me.

When I knew her in New York, Silvana was a precocious teenager with a black American boyfriend and an appetite for big-city nightlife. Now she had three young children (two of them silent infant twins), a big bear of a truck driver husband (Caucasian), and a new house in a nearby town. Silvana invited us to dinner. "Great," I said. "What can I bring?"

"Truffles!" was her reply.

After a joyful reunion, a tour of the new house, and an inventory of her oldest son's well-stocked toy box, we all sat down at the cramped kitchen table, her husband squeezing into a tight corner with a twin on each knee. Silvana prepared a fine meal of hand-chopped Carne Cruda, the subtle but full-flavored raw meat literally melting in our mouths, and delectable yellow homemade Tajarin tossed in butter. Both dishes were topped off with the finely shaved, earthy jewels I proudly pulled out of a brown paper bag. Her husband abstained (he doesn't like them), leaving all the more for the three of us.

"*Vai piano, eh,*" Silvana warned. "You know truffles are an aphrodisiac!"

"Perhaps you'd like to turn your trio of *bambini* into a quartet?" I joked, as I continued shaving generously over the women's plates.

I don't know if it was the truffles, or the wine, or simply the excitement of being in Italy again, but when we got back to the hotel that night, I invited Ivana up to my room. And she came. And she came back the next night and the one after that.

On the morning I finally had to leave, Ivana was alone at the front desk. We had a long and difficult good-bye, not knowing when—or even if—we would ever see each other again.

It was cold and gray and pouring rain.

The cooking class took place on schedule in a decrepit, depressing, and rather spooky old villa outside Florence, and afterward I returned to New York as planned. But I was not happy to be back. Though I had only been gone three weeks, the sound of New York English was grating; it was difficult to find a decent piece of salami or a good short espresso, and the fast-approaching holidays scared me. Fifteen months had passed since the terrorist attack, but the city still seemed in a muted state of shock. And after my time with Ivana and the rest of the gang in the cozy little village of Castiglione, I wasn't thrilled to return to self-imposed isolation amid the crowded canyons of glass and concrete. So, on the spur of the moment I went online and got a ticket back to Italy in time for New Year's.

· RECIPE #3 ·

INSALATA DI CARNE CRUDA BATTUTA AL COLTELLO

Hand-Chopped "Salad" of Raw Meat

THIS DISH APPEARS frequently throughout Piemonte, and some people still prepare it by hand (*battuta al coltello*). This takes a bit more work, but it makes a big difference. Machine-ground meat has a uniformly soft feel on the palate, but when it's hand-chopped, the pieces are slightly different in size and shape. The overall result of this un-uniformity is a lively, sensuously appealing texture. It dances around on your tongue and teasingly offers some fleshy resistance as it releases its delicate flavor.

Purists use only olive oil, salt, and pepper to dress *carne cruda;* others add lemon and a hint of fresh finely sliced garlic. Still others like to shave fresh truffles over the raw meat (as we did that night at Silvana's), in which case the subtle, earthy meat becomes a perfect springboard for the precious aromatic fungus.

INGREDIENTS

1 pound beef, cleaned of all fat,
    connective tissue, and sinew
    (see note below)
¼ cup of the very best extra-
    virgin olive oil you can get
    (I prefer Ligurian oil for this
    as it is especially light and
    aromatic)

Coarse salt and black pepper to
    taste
Optional: Lemon, fresh garlic
    (finely sliced), fresh truffles,
    or baby lettuces

1. Place the meat on a clean cutting board. Using a sharp chef's knife, slice the meat thinly against the grain. Cut the slices into strips and the strips into small cubes. Begin chopping the meat, going back and forth until it is as finely chopped as possible.

2. Transfer the chopped meat to a stainless-steel mixing bowl. Add enough olive oil to thoroughly moisten the meat, sprinkle with salt, and mix. Cover the meat in the bowl with plastic wrap and refrigerate for 20 to 30 minutes.

3. Remove the bowl from the fridge and taste: add more salt if necessary and fresh-ground pepper to taste. (If the meat has given off any liquid, blot it up from the bottom of the bowl with a paper towel.)

4. Serve. Here you have two basic options: if it is an informal dinner among friends, transfer the meat to a serving platter and garnish with the finely sliced garlic (go light!), lemon, lettuces, and/or white truffles. For a more formal presentation, use a metal ring to mound the meat onto individual dinner plates and garnish.

YIELD: SERVES 4 TO 6 AS AN APPETIZER.

———————

A NOTE ABOUT MEAT: Use the best quality, fresh, hormone-free lean beef you can get (numerous varieties of cattle, including *Razza Piemontese,* are now being raised in the United States). As it is to be eaten raw, make sure the meat is handled carefully: get it directly from a butcher, take it right home, and keep chilled until ready to use.

Meat from the leg would generally be good for this dish, but if you are using normal, fatty American beef, it might be better to use a piece from the filet, which, though a bit less flavorful, has less fat.

· RECIPE #4 ·

TAJARIN AI TARTUFI

Tagliatelle with White Truffles

TO MAKE *TAJARIN*, the dough is rolled out paper-thin and allowed to dry slightly. The sheet is then rolled up and the rolls sliced as thinly as possible with a sharp knife to form thin ribbons of pasta.

These days many people use a pasta machine to roll and cut the pasta, and this is perfectly fine, though it lacks the telltale irregularity of the hand-cut pasta.

Because Silvana knew I was bringing truffles, we had the very simplest of all preparations—just tossed with butter—leaving nothing whatsoever but the succulent mouth-feel of the pasta and inimitable flavor of the fresh truffles.

INGREDIENTS

3 ounces butter
Coarse salt
1 teaspoon extra-virgin olive oil

1 pound fresh tajarin (see note below)
1 fresh white truffle

1. Melt the butter in a large saucepan. Add a pinch of salt and the olive oil.

2. Cook the pasta in rapidly boiling salted water. When it is cooked (see note below), strain the pasta, reserving some of the water. Add the pasta to the saucepan and toss well (if the pasta is dry, add a bit more butter or a touch of the reserved water). Add more salt if necessary.

3. Transfer pasta to a serving platter or individual plates, shave fresh truffles over the top and serve.

YIELD: SERVES 4 AS A MAIN COURSE OR 6 TO 8 AS AN APPETIZER.

NOTE: Because of all the egg yolks, Piemontese pasta tends to be a bit heavier than normal fresh pasta made with whole eggs. You may, therefore, have to adjust the quantity of pasta based on the type of pasta you are using and how you are serving it—as an appetizer or main course, with meat sauce or wild mushrooms or butter and sage—as well as how hungry you are.

While actual cooking time will depend on the type of pasta being used, the pasta for these dishes should be thoroughly cooked. The Piemontese like their fresh pasta soft—not mushy and falling apart, but definitely not *al dente*—and this softness, along with the additional richness of butter, provides a perfect vehicle for the delicate, earthy flavor of truffles.

3 · GUESS WHO'S COMING TO DINNER

Tell me: what is the perfect wine for rapture? What is the correct fork?

—ANDREW SEAN GREER, *The Confessions of Max Tivoli*

PLUNGING DOWN THE steep winding road of Bussia Sottana in the premature darkness of a January evening, down we go as if into some big black hole in the earth, spiraling through sleeping vineyards cloaked in a soft wintry mist, past solitary farmhouses illuminated by the muffled glare of floodlights, shuttered tight, not a sign of life within. Off in the distance, the hills are nothing now but a faint black undulating outline against a dark gray backdrop; a lone pair of headlights moves determinedly across the darkness as if sailing through the sky. Night surrounds us like a thick woolen blanket, the profound silence punctured only by the sound of rubber tires on gravel. Just outside the thin shell of the plummeting car, the cool air carries with it the faint but promising perfume of wood smoke.

We pull off the road onto an even narrower driveway, pass under a high plaster arcade connecting two buildings, and stop at an iron fence.

The car doors slam shut with a thud against the nocturnal stillness. On cue, a huge German shepherd comes galloping up to the gate, scowling and annoyed, as if rudely awakened from some canine reverie. And when he sees me, he goes positively ballistic, jumping up in the air and barking out in a fierce basso profondo that suddenly transforms the evening's tranquillity into terror. He doesn't seem to like strangers.

"*Giù, Argo! Giù! Zitto! Stai buono! Giù!*" shouts Ivana.

"Nice dog," I quip, emboldened by the safety of the fence. But the sarcasm doesn't seem to come across in my Italian.

"Yes," Ivana replies, "although if I wasn't here he'd eat you alive." (I don't get any sarcasm in *her* Italian either). "Wait here a moment." She squeezes through the gate and, affectionately grabbing the furry monster around the neck, drags him across the yard into a shed and slams the door shut. I breathe a sigh of relief and enter.

This old farmhouse amid the sleeping vineyards is Ivana's home. She was born here and has lived here all her life, except for the six years she was married to a policeman in Alba. She left her husband three years ago and now is back here with her mother.

Despite the offhandedness of the invitation *("Are you free Sunday night? Would you like to come to my house for dinner?")* and my casual reply *("Sure, why not?"),* its significance is not lost on me. Nor on anyone else, for that matter: upon learning of the impending event, Renata teasingly refers to it as the *presentazione ufficiale*, the official presentation, and Bruna says simply, "*Ah, sì?*" slowly nodding her head with a strange and irritatingly knowing smile.

Castiglione is, after all, a small village and news travels remarkably fast; people notice whose car is parked where and when and who is spending time with whom. This "first dinner at the family home to meet the mother" appears to be a newsworthy item, and in no time it's all over town. Ivana and I let the inevitable innuendo wash right over us and just go about our business. But I was still a little apprehensive. I had heard stories about her mother's dislike of non-Piemontese, especially Southern Italians. Was an American tourist better or worse than an Italian from the south? I wondered. What about an American tourist of Southern Italian descent? Just to be on the safe side, I come bearing gifts, a purple potted cyclamen purchased that morning at the flower shop in Monforte.

Ivana pushes through the door, and we head up a narrow chilly staircase into the dining room where a big Persian cat is lying on a sofa and a small woman with close-cropped hair, smart tweed slacks, and a gray sweater is standing near the backwall, waiting for us. She reminds me of a snapping turtle without its shell.

"*Ciao, Mamma. Questo è Alan.*"

"*Buona sera, signora,*" I say in my best Italian. "*Molto piacere.* This is for you."

She awkwardly accepts the plant, and for a moment the entire top half of her body is hidden behind it.

"Here, Mamma, let's put that there," says Ivana, placing my offering firmly on top of the big TV. That diversion thus disposed of, we stand around trying to make polite conversation, but it's not easy. There are uncomfortable moments of silence during which you can practically hear the cat yawn, and I can't help wondering whether she has any idea what her daughter has been doing every night for the past week. Italian is a second language for both of us. Her first is Piemontese.

Familiar faces come to the rescue as Fabrizio and Marilena burst in carrying baby Pietro and tons of baby baggage: stroller, toys, blankets, and a bulging bag of diapers and backup clothes.

"*Guarda!*" exclaims Ivana. "*Che carino!*"

He is indeed adorable—at least the little bit I can see of him. He is swaddled head to toe in a dashing blue quilted Fila outfit with only the smallest patch of man-in-the-moon face exposed. He looks like an Italian prince, a Piemontese papoose, eyes tightly closed, full lips working away on his mother's imaginary nipple.

"*Ciao, Mamma. Ciao, Alan, come va?*" says Fabrizio matter-of-factly.

I had met Fabrizio and his wife when I was here the last time. In his official capacity as representative of the local wine *consorzio* (and in his unofficial capacity as Ivana's brother), he had taken me on a personal tour of the vineyards. Marilena was still pregnant then—very pregnant—and this was my first glimpse of the result.

All present and accounted for, we take our places at the large table in the middle of the dining room. Fabrizio at the head, Marilena on the outside near the sofa with the baby nearby in his stroller, and Ivana next to her. "Alan, you go sit inside beside Mamma."

Gingerly, I move around the table to my designated place. Just as I sit down, Mamma jumps up and leaves the room. (*She hates me*, I think.) After a momentary pause, everyone else springs into action: Ivana follows her mother back to the kitchen, Fabrizio opens a bottle of *spumante* and walks around the table pouring, Pietro wakes up and begins to howl. Marilena, extracting her child from his down cocoon, settles herself on the couch, lifts her shirt, and starts to nurse him.

Ivana and her mother return with salamis (a raw one and a cooked one) on a little wooden cutting board, a basket of bread, and a bowl of salad with pieces of soft *toma* cheese mixed in. Everyone lifts a glass in an obligatory *cin cin* and we begin. I am made to serve myself first. I gallantly try to pass the honor off to Ivana's mother sitting uneasily at my side, but no one will hear of it. *"Dai, Alan, forza. Prendi! Mangia!"*

Once we start eating things relax a bit. Marilena and Fabrizio talk about the baby and Ivana chats with her mom. Then the dialogues shift: Ivana talks with Marilena and Fabrizio with his mother; then Ivana and Fabrizio, and Marilena and her mother-in-law; sometimes everyone talks together. Even I jump in occasionally with a *"Sì?"* or *"Davverro!"* just to demonstrate that, though not an especially active participant, I am following the conversation.

Once the initial chitchat slows down, I even talk a little myself. I ask Fabrizio how local producers are doing after the disastrous 2002 vintage (heavy rain throughout the season and a preharvest hailstorm destroyed about 60 percent of the crop) and chat with Marilena over on the couch about her baby, who is still gorging himself at her breast. Mostly I try to converse with Ivana's mother, try with all the charm and wit I can muster to break through the thin layer of icy formality. I crack jokes in Italian; I even trot out my recitation of the days of the week in Piemontese, which everyone gets a big kick out of.

After the empty salad bowl and remaining salami are removed, Fabrizio goes to help his mother in the kitchen while Marilena, totally absorbed, continues nursing Pietro on the couch. Ivana smiles at me and reaches across the table to take my hand. *"Tutto okay?"* she asks softly, angling her head and raising her eyebrows. *"Sì, sì. Perfettamente okay,"* I reply.

Fabrizio returns with two long-handled copper pots and deposits them on the table. His mother follows with a plate of butter and a paper package. One by one, we pass our plates to the head as Fabrizio, stirring, ladles a large yellow mass of polenta onto each. "It's a little stiff, Mamma," he says. *"Eh beh, va bene lo stesso,"* she replies unperturbed. The other pot, containing pieces of pork, veal, and sausage braised in Barolo wine, is slid along the table so that everyone can help themselves.

Marilena adjusts her blouse, tucks the baby back in the stroller, and joins us at the table, and we begin to eat again. Fabrizio unwraps the paper, spoons Gorgonzola on top of his polenta, and mashes it in. So involved is he in this operation that he fails to notice his sister's persistent quizzical look. *"Fabri! Il vino?"* she finally blurts out. "Oh yes, I almost forgot." He spoons a big forkful of polenta into his mouth and disappears.

A few minutes later he returns cradling a label-less bottle and places it gently on the table. He gets a decanter down out of a big dark wooden heirloom cabinet. Then, standing at the head of the table, he carefully extracts the cork (it has no capsule), and we all watch in silence as he slowly lifts and tilts the bottle, letting the wine fall into the clear glass container. *"Questo vino,"* he announces solemnly, *"è Barolo settantuno."*

Barolo '71! Some time earlier, when I found out her age, I jokingly said to Ivana that I would one day like to share a bottle of wine from her birth year. As every lover of Barolo knows, 1971 is a legendary vintage and remaining bottles are highly coveted. They are also exorbitantly expensive if, that is, you can even find them.

This one, however, wasn't "found"; it came from the family cantina and was made by Ivana's father, who died in 1980. As is the custom, he set aside some extra bottles from the years in which his children were born. (Fabrizio has some from his year too, though unfortunately 1973 was a mediocre vintage). Birth wines are typically reserved for special occasions in the life of that person—first Holy Communion, graduations, marriage—and I feel quite honored to be sharing this one.

From the decanter, the wine is ceremoniously poured into large, thin-stemmed balloon glasses with a slightly flared lip. When we each have a glass I feel moved to make a silly toast: to baby Pietro (may 2003 be a better vintage than 2002), to Ivana for being born (and *che fortuna!* in 1971), to the memory of her deceased father, who made the wine, and to the continued good health of her mother—and then we taste.

As soon as it hits my palate, my eyes light up in anticipation: this is a wine to take note of. It is fantastic! A beautiful, Old World Barolo with the heightened complexity that only comes with long aging. The layered density of a mature wine mingles perfectly with the vivacity and tannic grip of youth (the bottle has obviously been well stored in the cantina in which it was born and the wine appears to be coasting comfortably at its peak). There are the wild, dark *frutti di bosco* flavors of black currants and ripe black cherries. There are mushrooms, truffles, even a hint of manure, and, on the finish, the characteristic (though increasingly rare) qualities of leather and tar.

This wine is not thick and jammy, the way so many wines today are made. In fact, in the glass it appears almost transparent, garnet (not velvety red) with a touch of orange, not unlike the outer skin of an onion. It is subtle and refined—it doesn't knock you over the head or get up in your face—yet it is incredibly intense. It begins in the nose with the smell of violets and moist, rich loamy soil; travels the full length of the palate from ripe fruit to tanned hide, and continues to evolve in the mouth even long after it's been swallowed.

This is what some might call a "wine of meditation," but that seems too passive a term—it is a more interactive experience, like a game of hide-and-seek: it keeps luring you back again and again to smell and taste and sip, and each time you do it is different; something else peeks out, something evocative and elusive. You go back and taste again to try and pinpoint it, but it is gone (or at least hiding in the shadows), replaced by yet something else.

For some time I am silently enthralled, intoxicated (though not at all drunk) by what's in the glass before me. Then I notice that everyone else is silent too; they're all looking at me and smiling.

"So," Fabrizio says after a while, "what do you think of the wine?"

A little embarrassed and self-conscious, I reply, "It's amazing, exceptional, really. It's . . . don't you agree?"

"Well, yes, I like it too. But I am biased. I am curious to know what you think."

I compose myself and try to give a more articulate summation of my impressions of the wine. I finish by saying: "Your father was quite a winemaker. Where did he get the grapes?"

"He grew them. They came from our vineyard. . . ."

· RECIPE #5 ·

BOCCONCINI AL BAROLO CON POLENTA

"Little Mouthfuls" Braised in Barolo with Polenta

BOCCONCINI MEANS "LITTLE mouthfuls," and what sets this preparation apart from a *stufato* (stew) or *spezzatino* (from the verb
spezzettare, to break up into small pieces) is the size of the pieces of
meat used to make it. It's a wonderful cold-weather dish, especially
with polenta. And the perfect wine to serve with it? Barolo, of course!

INGREDIENTS

2 tablespoons olive oil
½ pound sweet Italian sausage
¾ pound veal from shoulder,
    cleaned of excess fat and
    gristle and cut into 1½-inch
    cubes
¾ pound pork shoulder, cleaned
    of excess fat and gristle and
    cut into 1¼-inch cubes
Salt and pepper to taste
1 teaspoon dried oregano
1 tablespoon flour
1 small onion, peeled and diced
    (about 1 cup)
1 small carrot, peeled and diced
    (about ½ cup)
3 ribs celery, diced (about 1 cup)

1 branch fresh rosemary,
    wrapped in cheesecloth or
    tied with string (to prevent
    needles from falling off into
    stew)
1 clove garlic, crushed and
    coarsely chopped
2 cups Barolo or other red wine
6 ounces peeled, seeded, and
    diced tomato, fresh or
    canned (about ½ cup)
1 tablespoon tomato paste
1 cup veal stock or canned
    chicken broth
1 cup coarse cornmeal (see note
    below)
Water

1. Heat the oil in a large heavy-bottomed pan (preferably stainless
steel). Pierce the sausages, place them in the pan, and brown lightly
on both sides.

2. Season the meat with salt and pepper and the dried oregano, evenly dust with flour, patting to remove any excess, and add to the pan.

3. Cook the meat on a high flame to brown on all sides (the pan should be large enough to hold all the meat in one layer; if not, brown it in stages).

4. When the meat is golden brown, add the onions and stir. When the onions begin to turn transparent, add the carrots, celery, rosemary, and garlic. Remove the sausages from the pan. When they are cool enough to be handled, cut them into 1½-inch pieces and return them to the pan.

5. When the vegetables begin to caramelize, add the red wine (there should be enough to just barely cover the meat), scraping the bottom of the pot with a wooden spoon to lift up any caramelized bits. Continue cooking the wine on a high simmer until the alcohol is burned off and it begins to thicken (about 10 minutes). There should be about ½ cup of liquid left.

6. Add the diced tomatoes and tomato paste and stir. Add the chicken broth or veal stock. When it comes to a boil, lower the heat to a gentle simmer, cover, and cook for 45 minutes or until the meat is tender and the liquid has reduced to a homogeneous sauce (add more stock if necessary). Add salt and pepper to taste.

Serve with polenta and, if you like, Gorgonzola cheese.

YIELD: SERVES 4 TO 6 AS A MAIN COURSE.

────────────

NOTES: This entire dish can be made ahead of time. In fact, like most stews or braised dishes, it gets even better after a day or two. If you have leftovers, here's a great way to reheat them: rub the bottom of a baking pan or terra-cotta dish with olive oil or butter, cut the polenta into slices about ¾-inch thick, and line up along the bottom. Slightly warm the remaining *bocconcini*, adding a touch of water or stock if necessary, then pour over the polenta. Dot the top with pieces of Gorgonzola (optional) and bake in a preheated 350-degree oven for about 20 minutes.

For the polenta, I suggest you use a coarse, stone-ground cornmeal. (The one from Mulino Sobrino in La Morra is one of my local favorites; you might want to find out whether there is a stone mill near you.) If you want to use an imported cornmeal, try a *polenta taragna* from the Valle d'Aosta region, which has buckwheat flour mixed in with the cornmeal, giving it a wonderfully nutty flavor. Stone-ground cornmeal has a nice coarse texture and is more resilient than roller-ground cornmeal. It will take 30 to 40 minutes to prepare but is vastly superior to instant polenta, which tends to be flat, pasty, and insipid. A good coarse meal will also hold up better; you can make it a little bit ahead and keep it warm in a *bagna maria,* adding a bit of water now and again if necessary. It also makes great leftovers, fried, grilled, or baked.

The basic ratio is 1 part polenta to 4 parts water, but, as all corn meals are different, it is best to follow instructions on the package of the one you use.

Because it is inevitable that the polenta will bubble and spit as it cooks, use an extralarge, deep pot to prevent it from splattering all over the stove.

4 · FRANCESCO

FRANCESCO WAS A *contadino*. He was born in an old farmhouse beneath a vine-covered ridge called the Bussia, as were his father and his father's father and his great-grandfather. Francesco had cows, chickens, and pigs. He tended a vegetable garden and a grove of hazelnut trees. In the fall, he cut firewood and foraged for mushrooms and truffles. The rest of the time he tended the vines and made wine from his grapes. He learned how to do this from his father. The land had been in his family for generations.

In 1970, he married a girl, the daughter of a baker from a nearby low-lying village, and brought her to live in his house. Her name was Tiziana.

At first it wasn't easy for her. She missed the sounds of village life, the warmth of the big ovens, and the smells of yeast and freshly baked bread; she missed all the people who came into the store to buy their loaves and chat. Here it was cold and quiet, lonely. Here it was smelly and dirty; the dusty red earth was constantly turning up in every corner of the house, even under her fingernails. And the work was hard.

It was difficult at first to go from being a baker's daughter to a *contadino's* wife. But Tiziana got used to it. She loved Francesco and, though he didn't talk much, he was good to her. That first year they had a child, a daughter, and two years after that a son.

Life, though often hard, was good. Had they ever stopped to ask themselves, they would have said they were happy. But they never stopped; they lived their lives, they worked. Each season brought with it a different set of chores, and one followed the next in a never-ending flow. They didn't have much money, but the children had clothes to wear and the old house, with its thick stone walls, was cool in the summer and warm in the winter. There was food to eat and wine to drink. And the wine, moreover, was good. People practically fought to get Francesco's grapes. He gave them some, but most he kept himself to make his own wine. Whatever his family couldn't drink, he sold. But he made sure to put extra aside in the years his children were born.

Francesco didn't smoke, didn't play cards, and didn't drink—except for a glass of wine with his meal. He kept quiet and went about his business. What he *did* do, however, was dance. *Dio mio,* how he danced! The waltz, the polka, the mazurka—he could do them all beautifully, and he did, any chance he got, sailing around nonstop with Tiziana gracefully encircled in his arms. Because of the fine figure he cut on the dance floor and his habit of wearing leather gloves to Mass on Sunday, he developed a reputation as a gallant as well as a winemaker.

Ten years later, Francesco was dead. One night, he came home early complaining of a headache and numbness in his right arm. He lay down and, except for numerous visits to the hospital, never really got up again.

Overnight, the young wife became a widow, with two small children to raise, a farm to take care of, and the vineyard. Suitors came calling, but Tiziana wasn't interested. "How could any other man move into this house, Francesco's house, much less sleep in our bed? And how could any other man become father to these children, Francesco's children? Eh no," she said decisively, *"non va."* Neighboring winemakers came calling too. "Signora, sell me the vineyard. What are you going to do with it, a woman, and with two little kids? I'll give you a good price." But she turned them down too. Their "good price" was, she knew, a fraction of its real value. And besides, Francesco had always said, "Never, never lose the land."

Now Francesco was gone; she was alone, but his words stayed with her.

Luckily, they had managed to set aside a little money. One by one Tiziana sold off the cows. She had the pigs slaughtered and turned into salamis, sausages, and prosciutti. She kept the chickens for eggs (and an occasional dinner) and the little *orto* (for vegetables). And she kept the vineyard. She soon realized she couldn't work it herself, but she kept it nonetheless, renting it out to neighbors for a nominal charge just to prevent it from going fallow. *Never, never lose the land.*

Tiziana, who used to love to go dancing with her husband and socialize with their neighbors, turned inward. She hardly ever left the house after that. She didn't trust anyone. She minded her business and kept to herself. Life was even harder now, even lonelier, but they got by. . . .

Sitting around the table in the aftermath of dinner, sipping Barolo Chinato and chatting, I have a new respect for this tough little woman sitting next to me. Through quiet, stubborn determination she made it; she still has her home and the vineyard. Her children are grown. Fabrizio, now twenty-nine, has a wife and son of his own and an important position with the local wine *consorzio.*[1] And this year he has decided to take back the vineyard and work it himself, much as his father had when he was a child. Would I like to come and see?

"Yes," I reply, "I would indeed."

In the car on the way back up to Castiglione, Ivana flirts. "So, you liked my wine, eh?"

"Yes, I did."

"Really *eccezionale,* like everything from that year, no?"

"Yes, I'd say it was a stupendous vintage, for just about everything."

"Bravo. When we get home, perhaps I'll let you taste something else from 1971 that I think you will enjoy!"

5 · VINES AND WINES

GROWING GRAPES TO make wine is one of the oldest agricultural activities known to man and probably the most complicated. It is thought to have originated near the Caspian Sea in the fertile crescent of the Middle East, an area that is now primarily Muslim and almost completely devoid of alcohol. No one knows exactly when wine making began, but it was surely at the very beginning of civilization. There is archaeological evidence of wine making dating back to around 6000 B.C. during the Neolithic Period. Apparently, as soon as human beings began to settle in one place, they began to cultivate grapes.

The prehistoric techniques of grape growing and wine making were first codified by the Egyptians. Eventually the knowledge worked its way over to Greece, and there, in the temperate Mediterranean basin, the vine found an especially propitious home. Greek wine was highly prized by the Romans, who, in turn, planted vines throughout their expanding empire. Besides being the preferred beverage of the social elite, wine also became an essential fringe benefit for Roman soldiers and grape growing a means of bringing civilization and commerce to conquered territories. It was the Romans who first introduced grapevines to many of what are now the world-famous wine areas of Italy and France.[1]

Early "vineyards" bore little resemblance to those of today. They probably started out simply as wild vines transplanted to the base of trees; as the vine grew, it attached itself to the trunk and spread out along the branches—a kind of natural trellising system.[2] At some point, people realized they didn't really need the tree and could get the same effect by placing a tall stick in the ground near the vine. This permitted a much greater concentration of vines in a chosen, readily accessible area. It also exposed the grapes to more sunlight and better air circulation (less rot). Grapes grown in this manner were easier to harvest and must have tasted better than the ones grown in the trees (increased sun would have meant more sugar), but it also necessitated doing something with the long trailing tips of the vines. Early farmers quickly learned to attach cross sticks to the poles for the vines to run along; they also learned that vines lying on the ground would sink their claws into the earth, form roots, and start a new plant. Viticulture was born.

It is impossible to say which came first: did the discovery of fermented grape juice create the incentive for cultivating grapes, or did an excess of grapes grown as a food crop lead to the discovery of wine? Either way, the birth of wine was likely an accident. People were still primarily hunter-gatherers, and grapes were still first and foremost a foodstuff. But, because grapes contain an extremely high percentage of water, squeezing the liquid out of them was a no-brainer, especially when fresh drinking water was not always available. From there it is easy to make the leap to wine: at some point the juice fermented, people noticed the change that had taken place in the liquid (and in themselves when they drank it), and liked it.

The vineyard as we know it owes much of its existence to medieval monks, especially in France. The reasons are pretty obvious: They needed wine, both as dietary rations for their growing monastic communities and to celebrate the Eucharist with an increasing population of Christian believers (in the Catholic Church, sharing the communal chalice of wine with the laity during Mass was abolished during the time of the Plague). They had the land, thanks to nobles who donated it to the Church in exchange for indulgences, and the built-in labor force to cultivate it. Most important, the brothers had the brains. Monasteries were one of the few places where higher learning prevailed during the Dark Ages, so the monks were able not only to cultivate the vines but also to analyze, refine, and experiment with what they produced. They advanced viticultural techniques such as clonal selection, hybridization, and grafting and were among the first to recognize, appreciate, and explore the unique relationship of the vine to soil and sun.

It is this unique relationship that most sets grape growing apart from other forms of agriculture.

In many ways, the vine is like a factory. It has its physical plant and production lines, its R&D and marketing departments, and its quality control and expansion divisions. Production is based to a large extent on supply (the quality and amount of raw materials) and demand, as well as on external forces (such as weather). The vine's extensive root system anchors it deep in the earth absorbing water and minerals from the soil, while the leaves soak up the energy of the sun and transform it through photosynthesis. The vine itself shoots long branches up into the air and sends little tendrils out to wind around and grab on to anything they can, in order to support heavy clusters of fruit. The work of these complex manufacturing systems, and all the information, resources, and raw material they gather, go to one end: the grapes.

Grapes are a cultivated fruit, like oranges or apples or kiwis. But they're special. For one thing, there is an infinite variety of wine grapes (*Vitis vinifera*) as distinct from the relatively few species of table grapes. And they are much more sensitive to climatic conditions like sun, wind, and temperature than other types of fruit. Even nuances of a given place such as exposure and slope can have a major impact. Sinking their roots deep into the soil, vines seem to extract the essential qualities of the particular bit of earth they happen to find themselves in; they then translate these essential particulars into their fruit. That's why wine made from the same kind of grape in two nearby vineyards can be so different.

In this "translation," this process of extraction and transformation, grapes reflect the unique characteristics of a particular place and are transformed by them as well. They mutate and change significantly over time in response to their soil and surroundings. Transplant a grape from one country or climate to another and before long, it too will radically change in character; eventually it may even become a different grape altogether. (This has given rise to the complex science of ampelography, which attempts to trace and explain the history, origins, and relationships of various grape varieties.)

What most sets *Vitis vinifera* apart from other fruit is what is done with it: wine. Sure, other types of fruit can be fermented (apples, plums), but it doesn't come as easily or naturally. Grapes are naturally coated with a layer of yeast that, once the skin is broken, reacts chemically with the sugar in the juice and causes it to change to alcohol.[3] The raison d'être of this fruit is to be fermented and made into wine.

Wine can be a simple quaffing beverage (nothing wrong with that) and for millennia was an important source of calories for the masses. But it can also transcend the boundaries of mere thirst quencher, becoming a substance of unparalleled complexity and multiplicity that seems to ignite a whole host of sensory responses, of smells and tastes and experiences past.

All of the elements (minerals from the soil, water, sun) that went into the grape via the vine factory, the particular nuances of microclimate (slope, exposure, wind, etc.), plus the unique meteorological factors of a given growing season, turn up in some fashion in the final product—given, that is, a certain degree of skill on the part of the grower and winemaker.

And this too sets wine grapes apart. By definition, all cultivated fruits (and vegetables and grains) require the involvement of man, so there is always a built-in relationship between the cultivated and the cultivator. But never is it as close as it is with grapes. There is a deep and long-standing symbiosis between man and the vine, a mutual codependency, and many of their respective evolutionary journeys across the earth and through the millennia were undertaken together.[4]

I grew up in the Midwest. Wine was simply not a part of my life. Only much later when I started cooking did I begin drinking it and thinking about it and enjoying it. As a chef I had to consider wine as an important component of the dining experience and think about how its flavors and tannin and acidity interact with food. As a restaurateur, I had to taste many wines in order to select ones for my list and learn to recommend certain wines to my customers to match certain dishes and their personal preferences. Over the years I got more interested in and involved with wine above and beyond my cooking activities; I wrote about it, read about it, talked about it, and most important, tasted it whenever possible (sometimes, too, I just drank it).

So by the time I got to Italy I had already had some experience with wine. But it had taken place mostly in a contextual vacuum. I had little idea where wine came from or how it got there, of all that was behind the finished product—history and folklore, technology and tradition, skill and luck, nature and man—much less what resonance it might have for my own life. I had never really had a close encounter with grapes and vines on their home turf.

That was about to change.

6 · INTO THE VINEYARD

SEVERAL DAYS AFTER that first dinner at Ivana's, on a cold but sunny afternoon, I set out from Castiglione Falletto on foot up the twisting road toward Monforte. I turn right, as instructed, at the sign for Cascina Fontana, walk up the road a bit, and see a solitary figure down below in powder blue and red amid the bare brown vines. It's Fabrizio. He sees me and lifts his hand in greeting. I turn off the road and start down the steep path into the vineyard.

I have long known of Barolo, the wine. But Barolo is first and fore-most a place, a pretty little village with its prominent castle nestled among the hills. And now I'm actually here, standing in a steeply ter-raced vineyard of Barolo, practically spitting distance from the town of the same name, my feet tamping the thick, muddy earth. This, I think to myself, is the very vineyard where the grapes for that amazing wine I drank several nights before came from, the wine made by Ivana's fa-ther more than thirty years ago, in the year she was born!

Of course, I have visited vineyards before and taken many a walk-through with the resident winemaker to examine position and exposure, soil and trellising. But this is different. Walking through a vineyard (or any place for that matter) is different from standing still in one. And now, with feet firmly planted on the slanted earth,

I watch Fabrizio work, and my mind begins to drift. I think about language.

In Italian, the word for vine is *vite* (which also, oddly, means "screw"). It sounds a lot like—and for me is easily confused with—*vita* (life; also, oddly again, "waistline"). *Vite/vita*: two totally different words that just happen to sound extremely similar. But when you think about it, there seems to be a close underlying connection between them.

Life here is closely linked to the vine; most people earn their living from it and the seasonal rhythm of daily activities revolves around it. In a more general way, the gnarly vine seems emblematic of life itself, like a twisted umbilical cord uniting mankind to mother earth, providing sustenance. And wine, the fruit of the vine, has always been closely connected with—if not essential to—human life. Before the advent of modern plumbing and sanitation, wine was often a much safer, more accessible life-sustaining beverage than water and easier to store for longer periods. It also has the advantage of alcohol, which inhibits disease, lightens the spirits, and, some would say, makes life worth living. Wine is both lifeblood and (at least in the Christian tradition) Holy Blood, everyday thirst quencher and libation of sacrament and celebration.

There's also *potere* (power, potency) and *potare* (to prune). These words too are easily confused; one letter changes the whole meaning. But once again, their similarity seems more than mere coincidence.

Pruning, which is what we do that first day, *is* an act or demonstration of potency, of exerting power and control over the vines. And for Fabrizio, new father, who has just taken back this vineyard, pruning it for the first time must also carry with it the heady potency of reclaiming a family legacy and resuming a long-standing tradition. That first day I ask him how it feels to be working in his father's vineyard. "Good," he says, pausing momentarily. "It feels good."

Pruning is the first step of the annual growing cycle that, many months later, culminates in the harvest. By late October the grapes have all been picked; since then the bare vines have lain idle, resting lethargically, while their fruit takes on a life of its own in the cantina.

To prune, the farmer ventures back out into the vineyard, starting over, once again engaging himself in an age-old give-and-take collaboration with nature. He takes up his clippers and shakes up the vines, seizes control, chooses what to keep and what to discard. It is an extreme activity, perhaps even a little violent, in which the wild profusion of crackly brown vines is cut back to a single short stem.

Fabrizio shows me how to do it.[1]

"*Dunque*, here is the *vite*. You see all these branches? We have to select the nicest one, *quella più bella,* to leave for next year. Everything else goes. *Tak!*" He makes a few strategic cuts and a good half of the vine falls to the ground.

The nicest is the healthiest, the fullest (but not too woody) branch closest to the base of the vine. This is called the *capo a frutto* (fruiting cane in English), and it's the one that will be left to develop for the coming season. Sometimes the choice of the *capo a frutto* is clear; other times, when there is more than one desirable candidate to choose from (or none), it's more subjective, becoming a statement of personal preference, intuition, or viticultural prowess.

Once the *capo a frutto* is selected, everything else is cut off. On the lone remaining branch you count the buds, which are called *gemme*. Normally, you leave eight or nine—ten if it's a good one (in the olden days, when maximizing quantity rather than quality was the goal, it was not unusual to leave as many as fifteen) or five if it's not—and cut off the rest. (Fabrizio cautions to trim the *capo a frutto* before making the other cuts, just in case). Whenever possible, you also leave a short stub of a branch, one or two *gemme*'s worth, below the *capo a frutto*. This is called the *capo a legno* or *sperone* (*portao* in Piemontese, spur in English) and it's left in the hopes that it might generate a viable *capo a frutto* in the year or two to come. (As I was to learn later, one of the most important functions of the *portao*, especially on older vines, is to help regulate the overall height of the vine and keep the fruiting branch an optimal height off the ground relative to the slope and exposure of the vineyard.)

After pruning a number of vines while explaining what he's doing, Fabrizio hands me the clippers. "Now you try."

I begin slowly, timidly, under Fabrizio's watchful eye.

"This one?"

"Yes, okay."

"Ten *gemme?*"

"*Sì.*"

"*Uno, due, tre, quattro, cinque, sei, sette, otto, nove, dieci,*" I count aloud, moving my hand upward from one bud to the next. "Here?"

"*Sì, sì. Vai!*"

I trim the *capo a frutto;* snip. Then snip, snip, snip, tug, and the useless branches detach from the vine. We move on to the next vine, and then the next. Gradually I gain confidence, though I make no cut without Fabrizio's approval:

"And here, which one would you pick?"

"Umm, this one?"

"*No, quello non mi piace.* You see how it branches out farther up? That's no good. I say this one. It's thinner but definitely *più bella,* and the buds are nicely spaced." Snip!

After a while, Fabrizio decides to go home to get another pair of clippers. Left alone with the vines, I continue cautiously and tentatively on my own, carefully considering all the options before making my choice and, once I do, counting twice before making a cut. Before long he returns, and we resume together, now a two-clipper team.

We begin to move a little bit quicker, working side by side, from one vine to the next down the row. We talk less—I know a little better now what needs to be done—though we do stop from time to time to ponder the possibilities offered by a particular vine.

The sun is shining. It is cool, but we are working and I almost break a sweat. Off comes the outer layer, which I hang on a wooden post at the head of a row. Later the fog (*nebbia* in Italian, from which the nebbiolo grape gets its name) rolls in and dances visibly through the vineyards; up, down, one way, then another it goes, diaphanous, playful, mysterious, like the dancing spirit of a *masca,* a Piemontese witch. In a moment it engulfs us and you can't see twenty feet ahead. As the sun is hidden, a chill descends and I put my sweater back on, but we work on nonetheless. A nearby bell from an invisible church echoes through the thick mist and a moment later it is answered by another more distant one. Silence. Then the nearby bell rings again, and it hits me: I'm here working in a vineyard in Barolo. At this moment, there's no place I'd rather be. It is exhilarating, even magical, but in a quiet and under-stated way; strange as it may be, it also feels somehow completely natural.

Hours pass. Soon it is dusk, and we have to stop because it's getting hard to see. As we gather our things and put on our coats, we hear a car come to a stop on the gravel road at the top of the vineyard. A few moments later the figure of Ivana emerges through the haze, carefully picking her way down the path in full-length coat and high-heeled leather boots.

On the way back to Castiglione she teaches me a new phrase in Piemontese dialect: *Sun' ande a pué ant ij filagn.* I practice it over and over again, attempting to get the inflection just right, and later try it out on everyone I meet, much to their amusement:

"*I have gone to prune in the vineyards.*"

7 · BLENDINGS

THROUGHOUT EUROPE, AS the dominance of the Church and the nobility lessened (spurred by historical events such as the Enlighten-ment, the French Revolution, and the unification of Italy), land began to pass into the hands of the peasants, who had previously worked it.[1] This in itself was revolutionary; never before had the common man been able to actually own land, at least not on such a widespread scale.

Most of these new landowners were subsistence farmers. They had a small plot of land, grew vegetables and grapes, hunted and foraged, raised animals, and made wine primarily for their own consumption. Any excess was sold or traded at market. Most people hardly used money at all; in many cases, the only thing they really had to buy was salt (which, for that reason, was often highly taxed). To this day, the Barolo zone, like the Burgundy region of France, consists primarily of many small growers with tiny plots of land and a big sense of independence and individuality, quite different from other prominent wine areas like Bordeaux or Tuscany, which tend to have much larger estates.

Gradually, as transportation and shipping improved, commerce increased and geographical barriers were broken down. Wine became more of a marketable commodity and people who grew grapes were actually able to make some money at it. Most of them sold their fruit to consolidators (called *negoçiants* in France) or to larger, more commercial producers or cooperatives.[2] Some farmers, seeing an opportunity to add value to their crop, began to make wine and sell it in bulk; others took it a step further, bottling their wine themselves and putting their own label on it.

Fabrizio and Ivana's father, Francesco, was one of the first of the "new" (post–World War II) generation of Barolo producers to build a cantina in his home and bottle wine under his own label. Many people in the village, once they learn I've been working with Fabrizio, offer remembrances of Francesco:

"He was really a great winemaker, that Francesco. And quite a dancer too!"

"He and Tiziana were workhorses, always working rain or shine."

"What a cantina he made! At the time, nobody had a cantina like that in their home."

Tiziana tells stories of applying labels to the wine bottles by hand in the now unused portion of the house next door. But most of the cantina no longer exists; after Francesco died, whatever could be sold was, the rest crumbled away.

The vineyard too has seen better days. It has been rented out for the past twenty years to neighbors just to keep it from going fallow. Though it hasn't gone fallow, it hasn't been maintained particularly well either, not the way someone who owned it would take care of it, not the way Francesco used to keep it up.

Fabrizio remembers as a young kid hanging at his father's side while he worked in the vineyard. After secondary school, he attended agriculture school in Alba, majoring in viticulture, and went on to secure a job in the office of the *consorzio*. Now, however, he is ready: the lease on the vineyard is up, and he has decided (with his mother's blessing) not to renew it. He intends to take the land back and work it himself, in addition to his regular job.

This winter, Fabrizio, armed with information learned in school, practical advice from neighboring farmers, and a basic affinity for vines that is in his genes, dives into the vineyard. And I am there with him.

In a sense this little plot in the heart of Barolo has lain idle since his father died; now Fabrizio's going to transform it, to bring it back to life, to make it new and vital again. There are words for that in Italian: *rinascere* (to be reborn) or *rinnovare* (to renew). This is an exciting and worthwhile endeavor, and I am happy to be a part of it. Perhaps, in a way, I am even hoping to experience some sort of *rinascimento* myself.

I go back to help Fabrizio whenever I can—he can't afford to hire a helper—and in return he lets me do a lot more than most other people would. No one else, for example, would let a rank beginner like me prune the vines, actually make the critical choices about what to leave and what to cut.

I enjoy the work. I've always liked working with my hands. As a kid I used to love building with blocks and Tinkertoys and later on had a job restoring porcelain antiques, sculpting delicate fingers, noses, and flower petals out of paste. Later still, as a cook I enjoyed the tactile aspect of cooking—boning meat, cutting vegetables, filleting fish, mixing dough. But this is different. I'm outdoors for one thing, in the crisp air amid the beautiful rolling hills. And the vines are not just some inanimate material but a living, breathing piece of nature still attached to the earth.

I had long been a lover of wine, especially Barolo. But actually working with the vines begins to offer a different and much deeper perspective on the liquid in the bottle, like traveling back in time to experience a good friend or lover at key points in their past, long before they came into your life. Moreover, the work connects me to this new place in a way that I could not otherwise be connected.

After a day in the vineyard, I stop off at Renza's bar for a glass of Nebbiolo before going back to my room to wash up.

"Dove sei stato?" she asks.

"Ant ij filagn," I reply without missing a beat. *"Sun andé a pué!"*

I like having the dried soil stuck in the treads of my boots and the stray twig in my hair. Dressed in dirty blue jeans and a dusty down vest, I get a conspiratorial nod from other farmers in dirty blue jeans and dusty work clothes chugging by in their earth-encrusted tractors.

Before long, my hands start to become rough from handling the dried vines and clippers—but only a little. Bruna loves to tease me about the (relative) smoothness of my hands, turning my palms up and comparing them with hers. *"Guarda, guarda!"* she says with undisguised recrimination, "so soft, like a noble seigneur. Look at mine!" Despite many years in the kitchen, my hands *are* soft and smooth compared with hers. Hers (like those of most everyone else around here) have the unmistakable, ingrained hardness of someone who has grown up in the country working the land.

But there's something else too.

From the vineyard I can turn around at any moment and see Ivana's house, the house where she grew up, the house where I had dinner with her family only a short time ago. Then it was cloaked in darkness and mystery; now, from my slanted vantage point in the sharp light of day, it is clearly visible and, if not yet familiar, at least recognizable.

Everything starts to blend together. I spend days in the vineyard wrestling with the vines and nights in bed wrestling with Ivana. These two activities become pleasantly blurred in my mind so that when I handle the vines, I think of caressing her body. The new roughness of my hands seems only to enhance the softness and smoothness of her skin, which I survey thoroughly, intently, like an explorer mapping out unknown territory.

We are in the frenzy of a new romance, and our mutual exploration takes place mostly without words. We *feel* our way. This process is intensely passionate but not merely physical. When I am making love to her, the last thing I'm thinking about is wine. But later, in moments of calm reflection when I'm working up and down the rows of vines, I recall the escapades of the previous night and realize they are not dissimilar.

Wine is a sensual experience—smell, taste, how it feels in the mouth—beyond words. It is often extremely pleasurable but essentially indescribable. You can talk around it, try to use words as metaphors, as igniters to kindle a sense of recognition in someone you're trying to convey the experience to, but you can never quite replicate it or nail it down.

Our lovemaking too is beyond words. It is a form of communication, not easy to decipher perhaps, but nonetheless highly efficient and inexplicably profound. I still know so little about Ivana. But the connection is clearly there, and never more so than when we are making love: when I am inside her, connected to her like a root drawn down into the earth, I look into her eyes and feel I am looking through them into her innermost self. I feel that I *know* her in some essential if indescribable way. The rest—the words, the information—may come later, but it will only help articulate this initial deeper understanding.

8 · IVANA

RIGHT FROM THE start, it is clear Ivana is a unique individual, a person of starkly apparent contradictions. She grew up in an old farmhouse in the country surrounded by vineyards, daughter of a *contadino*, yet she has no interest in working the land or making wine, much less slaughtering the chickens that live in the little cage off the courtyard behind her house. She loves to eat but hates to cook and has a special fondness for sweets, especially her mother's chocolate pudding. And Ivana has a passion—bordering on obsession—for cleanliness and order.

Her mother and father always spoke Piemontese at home so Ivana understands the dialect completely but, from a very early age, refused to speak it. I ask her why:

"*Non mi piace,*" she says. "I find it ugly, the sounds. I prefer Italian or the way they talk down south. For me it is much more *simpatico.*"

In fact, with her dark hair and olive complexion, Ivana looks more like a southerner than a Piemontese. Her head is capped by a helmet of thick black hair with cascading bangs that spill over her left eye; underneath, her thin oval face narrows through a soft jaw coming to a point at her little recessed chin. Dark brows over large brown eyes with fine creases in the lower corners and a delicately angular nose give her a certain avian aspect, like an owl or a raven or, better yet, like one of the blackbirds that come to nibble the dried grapes in the vineyard. (Ivana was born on the last day of January during what is known as the *Giorni della Merla,* the Days of the Blackbird.)

The thinness of her face continues down through her shoulders and arms, slender hips, and flat stomach, blossoming out midway to her prized feature, her round voluptuous behind. Like many Italian women, Ivana is honestly matter-of-fact about her physical attributes: she feels her breasts are small, but she is unequivocally proud of her *"bel culo brasiliano"* and likes to wear tight pants or short skirts to show it off.

Ivana is beautiful. With her lithe body, elongated oval face, and dark coloring, Ivana has the mysterious, exotic allure of a Modigliani, but without the aloofness. She is a lighthearted person, *allegra;* she loves to laugh, enjoys Disney cartoons and silly stand-up comedy, and likes to be around other people.

Ivana pays regular visits to her aesthetician in Alba for tanning and waxing and to her hairdresser in Barolo for highlights and a cut. She wears no makeup but is fond of jewelry and sports a lot of it: two earrings in each ear, a diamond nose stud, a black pearl necklace on a white-gold chain (from me), several bracelets on each wrist and several rings on each finger—one, she is pleased to confess, from each of her past boyfriends.

And she has almost as many names as pieces of jewelry: Ivana Margherita Antonia Teresa Mascarello.

While Ivana may be atypical, she shares one characteristic with her fellow Piemontese; she is a *testona,* a hardhead. She is stubborn and proud and opinionated. She is fiercely loyal but merciless if she feels she has been betrayed or taken advantage of.

And this raises yet another apparent contradiction: underneath that tough, sprightly outer shell, she is tender, sensitive, and perhaps even a little bit vulnerable.

Like her father, Ivana loves to dance. From the age of fourteen until she got married at twenty-two, she and her friends made a regular tour of the area discotheques, often driving as much as an hour to get to the right place on the right day and dance until dawn. Sometimes she still goes out dancing with Renata to let loose, to get blinded by the flashing lights, and to be swept away by the pounding music. At the *discoteca*, time stops, little nuisances and larger problems fade away. Ivana's fluid body becomes one with the song while she quietly sings along to herself—quietly, because she is admittedly tone-deaf and unable to carry a tune.

Ivana and I are about as different as two people can be:

She is a product of the vine-covered hills of rural Italy, I spent most of my life in the big cities of America. I have a college degree, read voraciously (when not working sixteen hours a day in restaurants), and have a tendency toward the creative and the intellectual; she rarely reads, and if she does, it is likely to be a popular mystery or illustrated book of jokes at which she sometimes laughs out loud. Though as a child, it was important for her to do well in school so as not to be shamed and looked down upon by her teachers and peers, she is decidedly unacademic; she would much rather dance than study, laugh at a silly joke than read a serious book, and bask in the warm glow of human companionship than engage in some solitary creative pursuit. After finishing high school, Ivana enrolled in a sommelier course but, due to outside events, never managed to finish. When she writes, it is practically illegible (sometimes even to herself), and when she draws it is most often stick figures outside a house with shining sun. While I have traveled extensively throughout the United States and Europe, Ivana has taken few trips out of her home territory of Piemonte.

Ivana and I are about as different as two people can be, as different as the extreme opposites of New York City and Castiglione Falletto, as different as the noble European *Vitis vinifera* and the wild American *Vitis lambrusca*. Yet we are drawn to one another, like the two opposing forces of a magnetic pull. I wonder—is it precisely this difference that forms the basis of attraction? Should we try to resist the pull or just give in fully to its tight embrace, like surrendering to gravity? And if we do give in (as give in we must), where will it lead?

· RECIPE #6 ·

BONÈT DELLA NONNA

Grandmother's Classic Chocolate Pudding

INGREDIENTS

½ cup sugar, plus ¾ cup sugar
8 extra-large eggs
3 ounces (approximately ¾ cup)
   sifted bitter dark cocoa
   powder (see note below)

2 tablespoons rum
¼ pound amaretti *cookies*
1 quart milk

1. Place ½ cup sugar in a heavy-bottomed sauce pot. Mix in about 1 tablespoon of water (just enough to dissolve the sugar), place over medium heat, and cook until a golden brown caramel is formed.

2. Pour the caramel into a 1½-quart (9 × 5 × 3-inch) loaf pan, and tilt it back and forth in order to completely and evenly cover the bottom. Set aside.

3. Crack the eggs into a large mixing bowl, add the remaining ¾ cup of sugar, and whisk. Add the powdered chocolate and the rum, and whisk to combine well.

4. Break the *amaretti* cookies into a saucepan; add the milk and heat over low flame until the cookies dissolve (about 15 minutes). Turn off the heat and let stand until lukewarm (10 minutes).

5. Preheat the oven to 350 degrees.

6. Add half the warm milk mixture to the chocolate and eggs, one ladle at a time, mixing well to incorporate each time. Then add the remaining milk and whisk well.

7. Pour cocoa mixture into the loaf pan, cover with foil, and place the loaf pan in a small roasting pan with sides 3 to 4 inches high. Add

enough warm water to reach halfway up the sides of the loaf pan and place the whole thing in the oven.

8. Cook the Bonèt for 45 minutes or until a wooden toothpick inserted into the center comes out clean.

9. Let cool. Slide a thin knife along the sides, place a platter over the loaf pan, and invert. Slice, spoon some of the caramel over the top, and serve.

YIELD: SERVES 8 NORMAL PEOPLE OR 2 IVANAS.

———————

A NOTE ABOUT CHOCOLATE: As this dessert is all about chocolate, the one you use will make a big difference in the final product. I prefer to use a dark, bitter powdered chocolate such as Valrhona from France, Cuba Venchi from Piemonte, or Scharffen Berger from America.

9 · TONGUES

DAYS PASS. THE New Year loses some of its newness while Fabrizio and I finish pruning the vineyard.

And then it snows.

I wake up one morning, look out the window, and find a completely altered landscape. Everything, from nearby terra-cotta-roofed houses to distant hills, is covered in a thick layer of white. Familiar buildings are hidden or disguised, hard edges rounded. And it's still falling, the fat flakes floating down in slow motion or blown harshly sideways by a sudden gust of wind. In response, almost everything comes to a standstill, and I am happily stranded all day long in my little fortified hilltop refuge.

At night, it finally stops snowing, and Ivana, bravely navigating the steep, slippery roads, arrives. The thick coating of snow only intensifies the normal hushed quiet of Castiglione; you can barely hear the dulled clanking of the campanile outside our room or the muffled sounds of lovemaking within. The following day, the sun returns and melts the snow off the roads. And that afternoon I venture out to have a look around.

After my time in the vineyard, I view the surrounding landscape differently, notice things I would not have noticed before. Now I can quickly pick out the pruned vines from the unpruned ones, even from a distance. (It is still early in the season and pruning has just begun. Many people prefer to wait a bit, allowing the vines a longer rest after the trauma of harvest. Others, like Fabrizio, have regular full-time jobs and need to get the work done when they can, the sooner the better. Most everyone waits at least until mid-December when the leaves have dropped from the vines. In any case, pruning should be finished by mid-March.)

The pruned vines, their single skinny branch sprouting from the thick gnarly base, appear naked and exposed against the stark white snow, even a bit awkward and self-conscious, like a group of shivering young boys at swim class. Distant villages—Barolo, La Morra, Serralunga—stand out too, along with the stray farmhouse or *ciabot* (shed), and the sharp winter sun bouncing off the whitened ripples of land makes everything sparkle.

It is these ripples that give the area its name—Le Langhe—which, by popular consensus, comes from a bastardized version of the Italian word *lingue,* the French *langues,* or the Piemontese *lenghe*: tongues.

I like the idea of being in a place called the Tongues and can think of several reasons why *I* might refer to it by that name. I didn't understand why anyone else would though. For most Americans the word conjures an image of something red and moist and fleshy, almost tropical, which this landscape is definitely not. But after my first visit to a local butcher shop, I got it. Sticking out at me from behind a glass case was a calf's tongue; thick at the back (where it was attached) and flat at the base, it starts out high and wide, then rolls down gently to the thin, tapered tip.

·Whoever first came up with the name—some anonymous Piemontese *contadino,* no doubt—did so with perfect poetic aptness and accuracy. Everywhere you look, "tongues" of earth covered with vines or trees or houses are lolling off into the distance. From certain vantage points, especially on misty days, these ripples, lined up one behind the other, resemble sand dunes in the desert or wave breaks on a wide ocean surf. As geological features go, these lateral hills are soft and round and supple, carved by nature, molded by time and man. They start out high and wide and gradually roll down lengthwise before disappearing into the valley floor. Often there is a road down the central spine with a house or cantina and vineyards spilling down the sides, which are often quite steep.

Hillsides are good for vines, exposing them to precious sunlight the way seats at a movie theater do the screen, and providing good circulation of air, which helps prevent rot. In fact, the slanted hillsides in between two tongues often resemble an amphitheater or a scallop shell fan (conca). The plots on the shoulders of the slopes are usually better than ones at the top as they get optimum exposure and ventilation without feeling the whip of the wind.

Vines can be planted either *ritocchino*, straight up and down, or *girapoggio*, horizontally along the contour of the hillside. The choice has something to do with harvesting logistics; if the slope is too steep, it can sometimes be quite difficult to work up and down the rows. But mostly it's a question of exposure. Which way, given the slant of the hill and the direction it faces, will give the grapes the most sunlight? Generally, south or southwest exposure is best because it gets sun from morning till dusk, though in the hilly region of "the Tongues," a lot depends on what's around it—a vineyard might have perfect southwest exposure but if there's another hill right in front of it, sunlight will be curtailed and the plot less desirable.

Farmers know, both by firsthand experience and generations of trial and error, what are the best locations for grapes and how best to plant them. The best sites in the zone are reserved for nebbiolo for Barolo while the less desirable ones are planted to dolcetto, barbera, or cabernet or white grapes such as favorita, sauvignon, or chardonnay. Within the Barolo zone it's forbidden to use nebbiolo grown on north-facing slopes to make Barolo, the collective assumption being that grapes grown on hillsides with such inferior exposure could never produce a wine worthy of the appellation.[1]

While these giant "tongues" are ideal environments for growing grapes, they were formed millions of years ago, long before vines found their way here. In the Eocene epoch about 40 million years ago, the southern supercontinent known as Gondwanaland collided with the northern supercontinent Eurasia, forming the Alps (as well as a number of other major mountain ranges, such as the Himalayas). This monumental event caused a ripple effect to the south, creating bumps and buckles in the earth. Besides the tall mountain ranges and sub-alpine foothills, a series of inland pockets were formed, which filled with water; the territory that is now Piemonte was once almost en-tirely submerged. Underwater shifting and sedimentation of sandstone and clay during the Miocene epoch (24 or so million years ago), along with deposits of ancient shell and petrified marine life, form the basis of present-day soil composition. If you were to take a deep cross sec-tion of earth, you'd be able to see numerous strata of material caused by silting and shifting over the millennia.

During the Pliocene epoch, the ripples and bumps in the earth were thrust up even farther through continuous pressure of the south-ern tectonic plate, lifting many areas up above water. A protracted pe-riod of intense rain during the early Quaternary period led to extensive erosion, which washed away much of the soft topsoil from the upper peaks. Then, during the most recent Ice Age of the Pleistocene epoch (1.6 million to 10,000 years ago), this abundant water froze. When it later melted, water rushing down from higher altitudes carved even deeper crevices and indentations in the elongated hills on its way down to low-lying basins.

After this last thaw, nature's sculpting of the Langhe landscape was basically complete, ready and waiting for *Vitis vinifera* to arrive.

Due to this long and complex series of geologic events, the odd dis-placement of alluvial, mineral-rich soils, and the numerous exposures and weather patterns caused by the rippled tongues of earth, few wine regions can boast the compressed multiplicity and diversity of microcli-mates that the Langhe has. It changes radically from zone to zone within the area and from one vineyard to the next within a given zone.[2] Even within a single small vineyard, soil composition can vary tremendously, producing very different results. Certain vines or parts of a row may ma-ture earlier or be of noticeably higher quality.

Between the hillsides of ordered vines are crevices filled with a dense profusion of growth. In some cases these wild spaces are well defined and restricted, while in others they are more extensive. Because of these crevices, accurately gauging distance can be tricky, as I found out one dreary day when I went out for a walk.

I set out from the back of Castiglione behind the church, down through the steep vineyards and up the twisting road to the sleepy hamlet of Perno. I was looking for the *funtanin,* a fresh-water spring coming out of the ground somewhere between Perno and the Bussia. As I reached the lovely, deserted (but not unconsecrated) tiny hilltop Chapel of Santo Stefano with its overgrown cemetery next door, it started to drizzle. I cut down across the vineyards, heading west toward the big white cantina of Aldo Conterno.[3] When I reached the end of the vines, I entered into thick, wild forest. The temperature dropped, it got ominously dark, and I soon came upon a deep, impassable ravine filled with tall trees and dense undergrowth.[4] I cautiously peered over the edge; the bottom wasn't even visible, and I imagined falling in, disappearing, and no one being able to find me. With mounting desperation, I circled around and around in the now steadily falling rain looking for a safe way across. I didn't want to turn around and go back the way I came; it would take too long, and besides, I knew there *had* to be a crossing. And indeed there was, it just took me a long time to find it.

When I finally came out on the other side, it was nearly twilight. Heart pounding, I made my way up to the road and back toward Castiglione like someone returning from another world. I got to my room, shed my filthy wet clothes, and stumbled into a nice hot shower, happy to be alive but pissed not to have found the *funtanin.*

Unlike my first visit in November when I was just passing through on my way somewhere else, this time I had come precisely to be here in Castiglione Falletto, here in the land of the Tongues—and to see Ivana. I thought ten days would be sufficient, but they flew by. I extended my trip by another six, which cost me a stiff penalty. But it was worth it. Ivana came to spend every night with me in the hotel, and each was like a gift, unexpected but gratefully accepted and gladly reciprocated.

Once again, the time came to leave. And once again, I got home only to begin planning the next trip.

· RECIPE #7 ·

LINGUA AL VERDE

Calf's Tongue in Green Sauce

INGREDIENTS

1 calf's tongue (see note below)

1 teaspoon coarse salt

2 ribs celery, cut in half

1 small onion, peeled and quartered

1 carrot, washed and quartered

1 bay leaf

½ teaspoon black peppercorns (optional)

1 tablespoon plus 1 teaspoon white wine vinegar

2 anchovies (4 fillets), rinsed, patted dry, and any fins or bones removed

1 teaspoon capers

1 clove garlic, finely chopped

2 cups (packed) flat-leaf Italian parsley leaves (no stems), washed and patted dry

1 cup extra-virgin olive oil

1 piece of day-old Italian-style bread

¼ cup milk

I. COOK THE CALF'S TONGUE

1. Place the calf's tongue in a medium-sized stockpot and cover with cold water. Add half the salt, the celery, onion, carrot, bay leaf, peppercorns, and 1 tablespoon of the vinegar.

2. Bring to a boil, then cover, lower heat to a gentle simmer, and cook for about 40 minutes or until a knife inserted into the center goes in easily.

3. Turn off the heat and let the tongue cool in the liquid. When it has cooled down enough to handle, remove it from the pot and peel off the outer skin. Place the tongue in a plastic storage container and add enough of the cooking liquid to cover (discard the rest). Cover the container and refrigerate until thoroughly cool. (May be prepared a day or two ahead.)

II. PREPARE THE GREEN SAUCE

1. Place the anchovies, capers, and half the garlic in the bowl of a food processor fitted with the chopping attachment, and pulse to chop finely.

2. Add the parsley and pulse. Add the olive oil little by little, processing in between to mix well.

3. When about half the oil has been added, taste the mixture; add more of the garlic, if necessary, and the remaining wine vinegar, then continue adding the rest of the oil.

4. While the sauce is "resting," cut or break the bread into large pieces and put it in a small bowl. Pour the milk over the bread and toss to mix well (if it is too dry, add a bit more milk). When the bread has absorbed all the milk and has begun to get soft, squeeze it firmly in your hands to remove the excess liquid. You should have about two tablespoons of moist, compressed bread.

5. Add half the bread to the green sauce and pulse to mix thoroughly. The sauce should be bright green and homogeneous, with faint little white specks. If it is not homogeneous, that is, if the oil separates from the sauce, add a bit more of the bread.

6. Adjust the salt, vinegar, garlic, and anchovy, if necessary. Put the sauce in a bowl, cover, and refrigerate until ready to use. (The sauce is best used within twenty-four hours as the green color will begin to fade.)

III. MARINATE

1. Remove the cooked tongue from the liquid and slice crosswise as thinly as possible. (Store any leftover tongue in the liquid.)

2. Arrange the slices on individual plates or in a shallow ceramic dish and spoon the sauce over to cover.

YIELD: AN AVERAGE-SIZE CALF'S TONGUE WILL SERVE 4 TO 6 PEOPLE AS AN APPETIZER.

NOTE: Calf's tongue is available in most quality butcher shops. If you don't want to invest in a whole tongue, you may be able to purchase slices of precooked tongue in specialty delicatessens.

10 · TYING THE KNOT

THE NEXT TRIP turned out to be about five weeks later. Ivana's mother has finally agreed to have a very serious operation. Her open-heart surgery is scheduled (ironically) for Valentine's Day, and Tiziana is terrified. Ivana is terrified too, and it falls upon her to make most of the decisions, schedule tests, talk to doctors, and try to explain everything to her mother. I want to go and help out in whatever way I can. Besides, there is little to keep me in New York.

I arrive several days after the operation took place, and that same evening Ivana and I go to visit her mom in the hospital. Everything went well. She is out of intensive care now and, though even frailer than before, is able to get up and walk around on her own. The next day she is to be transferred to a rehabilitation center where she will stay for two weeks, though she would rather go straight home to the Bussia and see her cat. I feel bad for her, but in a way I'm glad. She really needs to take it easy; the rehabilitation center has a great reputation, and she will surely benefit from the professional care and attention she will receive there. Plus, it gives me and Ivana a little more time together.

Hotel Le Torri is closed for its annual two-week vacation, but Silvy has graciously offered to let me stay there anyway. She has turned on the heat in the little room on the ground floor, made up the bed (all the others are stripped down), and given me a key to the padlock on the gate.

I like staying in the empty hotel. In the dead of winter, sleepy Castiglione is even sleepier, and I'm beginning to feel more like a resident than a visitor. People are getting used to seeing me around too; many of them don't even realize I've been away.

It is cold here, even colder than it was in January, but it still doesn't have the harshness of winter in New York. During the day the sun shines, making it comfortable to spend time outdoors. I hike the muddy trails through the vineyards, walk the deserted streets of Monforte Vecchio, and jog along the asphalt road at the base of Castiglione.

And I go back to help Fabrizio in the vineyard.

There too I have the same warm feeling of returning to a familiar place. But what is it called? (In Italy, every vineyard—and in a rural area like this one, every distinguishable little patch of ground—has a name.)

"Le Munie," replies Fabrizio when I ask him the name. "It has always been called Le Munie. It means 'the sisters' in our dialect. Sisters as in nuns. There used to be a convent here somewhere. No one knows exactly where, but there was certainly something here. You can feel it, something sacred."

We go to work, starting with *scalusé*, a general putting to rights of the vineyard before the next really important step.

Fabrizio goes up and down the rows checking for loose, wobbly wooden poles. When he finds one, he pulls it out and, using a short-handled scythe-like blade that dangles from his belt, hacks rotten wood off the bottom into a point. Then he breaks out a heavy, three-foot-long iron tool; at one end, the tip is pointed like an elongated pyramid about 8 inches long and 4 × 4 inches at the base, and at the other end is a flat round knob. To start a new hole or make an existing one wider or deeper, he plunges the pointy end into the ground and rotates the tool to form the hole. When the wood pole is thrust back into the hole, he bangs the top of it with the flat side of the iron pyramid and uses the knobby end (along with his feet) to tamp down the earth around it.

Rotting wood, shifting earth, blowing wind, and the weight of the vines mean these wooden poles must be checked, fixed, and in some cases replaced each year. I ask Fabrizio what he thinks of the newer cement or metal posts often seen around here:

"They're good, they're sturdy, and they last forever (unless you bump into them with a tractor); with them you don't have to do all this work. But they're expensive. Some day I'll change everything. *Pazienza!*"

Once all the poles in a row have been stabilized, Fabrizio pulls taut the three metal wires—low, medium, and high—that run the length of the rows and reattaches them to the posts at the head of each row. I follow behind him securing the metal wires to the wooden poles with brown paper–covered wire twist ties. The wires must be reasonably taut and secure in order to support the weight of the vines and grapes that will come.

When we have done all this, when the poles are stable and the wires taut and secure, we can begin the real work: *tòrse*, a Piemontese word used to describe the process in which the fruiting cane is bent (*piegato*) and tied (*legato*) to a wire. This is a major, critical step in the process of wine-grape growing as it is done in Piemonte, where a Guyot-type system of trellising is traditionally used. The *capo a frutto,* the sole remaining branch after pruning, must be bent horizontal and attached to the lowest of the wires so that, when the buds blossom, they send their shoots straight up.

This is more complicated than it sounds. The vine is still basically in hibernation. While some branches are supple and bend gracefully like willows, others are dry, stiff, and brittle, and must be coaxed, gently but firmly, to bend. Each branch is different and has to be approached with an open mind and a sensitive hand.

It doesn't really matter which way you bend it, though some vines (like people who are right-handed or left-handed) do seem to have a natural inclination toward one side or the other. Many traditionalists feel that they must all go in the same direction (which does make later work from one vine to another down a row somewhat easier) while others, like Fabrizio, don't think it makes much difference.[1] The most important thing is to try to maximize the linear space along a row of vines by completely filling the horizontal wires with bent branches—without, however, overlapping them—thus exposing as many buds, and the branches that will later come from them, to sun and circulating air.

Here's how it's done: the right hand grasps the tip of the vine and begins to bend it while the left hand works lower down on the branch near where it meets the base of the vine. The left hand acts like a kind of fulcrum. As you bend, you test the stiffness or willingness of the particular vine. Starting near the juncture and working your way up, you press or squeeze with the thumb and index finger to soften the branch at the precise point of most tension. It's kind of like a trainer warming up the tight muscles of an athlete; stretch, massage, coax, cajole, and bend. The branch has a dry outer skin that often cracks open when you bend it, revealing the supple pale green hiding beneath. Sometimes you can actually feel the tight filaments of the branch give way, and occasionally you may even hear a kind of pop as the branch relaxes and surrenders.

When the branch is bent to the lowest wire, you secure the tip with a wire tie *(legaccio)* and place another tie farther back where the branch first meets the wire. Instead of wire ties, some people use plastic zip ties or string. There's also a nifty handheld tool that ties and cuts a piece of string around the branch in the twinkling of an eye, and some people still use the traditional tie (called a *gurat*) made from the supple tips of willow branches.

Whatever type of tie is used, the important thing is not to break the branch as you bend it. The easiest place to break it is right at the joint, though it's also possible to crack it mid-branch, especially if a person is rough and careless. *Tòrse* requires a firm yet gentle touch and a strong will. For this reason, women are not only permitted but often encouraged to perform this task; some growers, in fact, will only let women do it.

Generally, it's best to wait for a period of lots of fog or humidity or until after a rain when the branches are the most supple. But you can't always wait for it to rain, and, in any case, breaking some branches is inevitable. Sometimes they are just too dry and brittle. Occasionally, just as you reach the point where the branch has finally submitted, met the wire, and is about to be tied off, you hear a pop and the branch breaks off in your hands. Inevitable though it may be, it is a sad moment. The vine, deprived of its buds, is basically useless and will produce no grapes for at least a year. *Che peccato!*

Thankfully, it doesn't happen very often.

Once *tòrse* is finished, another level of order has been imposed on the vineyard: the stiffly arching, tied branches resemble soldiers in formation preparing to march into battle or a bunch of little catapults set to launch their pebblelike buds up at the blue sky. And before too long, this is just what they will do.

Even in the dead of winter you can feel the impending approach of spring.

I went to Ash Wednesday Mass in the pristine little church of Castiglione. I had been inside only once before, all alone, one afternoon when the sun shining through the stained glass windows cast colored reflections on the frescoed walls.

Now, at eight forty-five in the evening, the place was packed. It was a lovely service, simple but sincere, a far cry from the fancy liturgy and clouds of incense at my church back home. The choir consisted of eight villagers bundled up in their winter coats standing in a semicircle behind the altar belting out songs to the tentative accompaniment of a little organ. There was no printed program, but it wasn't really necessary: everyone in attendance knew what to do.

Though not religious in the formal sense, I have always loved the period of Lent, the forty days (*Quaresima*) between Ash Wednesday and Easter. Like the Jewish Passover, it stems from the ancient, pagan, quintessentially human rites of spring. It is a time of darkness before light (the short days of winter changing to the longer, warmer, sunnier days of spring and summer), a time of quiet introspection, self-deprivation, and, ultimately, rebirth; a celebration of change.

The whole Passion of Christ—the crucifixion, resurrection, and ascension—is in a way a metaphor (literal or figurative, depending on your belief), a vehicle for experiencing this transformation of the natural world and infusing it with life, humanity, and significance. The Passion also mirrors changes taking place in the vineyard. The vines have been cut and the remaining branches tied down to the crosswire; the buds on the still-brown *capo a frutto* are plump now and just about ready to burst open, sending forth their shoots. At some point during this time the *Giorni del Pianto* will take place, the "Days of Crying," when the vines emit "tears" of transparent sap from the cuts made months before.

Walking through the vineyards in the pregnant silence of late winter, you have the feeling that something significant is about to happen.

This Lent also seems to reflect changes taking place in my own life. This year I had chosen to give up smoking, which, since I was only an occasional smoker, turned out to be a pretty easy penance. But because I once again left Castiglione for New York just after Ash Wednesday, I also gave up something much more cherished—sex. During this third visit, Ivana and I had talked about us: we decided that *siamo in due,* that we were a couple, and therefore would not see other people.

This was a mutual decision and a logical one, based on the simple acknowledgment of our feelings for one another. But it created some problems as well: I lived in New York City, U.S.A.; Ivana in La Bussia Sottana, Comune di Monforte d'Alba, Provincia di Cuneo, Italia. Neither one of us was especially excited about the prospect of an extended long-distance relationship. Not to mention the fact that phone bills (when I was home) and airfare expenses (when I was not) were adding up while the cost of maintaining my empty apartment in Manhattan was definitely *not* going down.

There was really no question of Ivana coming to the States. She needed to be in Italy to look after her mother. Besides, her English was just a tad better than my Piemontese. And what would she do in New York anyway? Why bring her to a place that I wanted to get away from? I liked being in Italy: Castiglione represented a nice—if rather extreme—change of pace. I enjoyed the cultural immersion of spending time in a small village in an unusual region of Italy. It offered the opportunity to significantly improve my Italian, maybe even learn some dialect. Plus there was the vineyard. . . .

The situation called for a creative solution and drastic action.

I had an idea.

PART II

SPRING
CLEANING

II · RITES OF SPRING

APRIL IN MANHATTAN. There may be no song to immortalize it as there is for Paris, but it's a fantastic place to be. Everywhere you look trees—anonymous or invisible the rest of the year—explode in flagrant fragrant flower. Technicolor tulips and forsythia seem to pop right up out of cracks in the cement, and a bevy of breathtakingly beautiful women reappear like a flock of exotic migrant birds, shed their heavy winter coats, and prance bare-midriffed and tan-shouldered through the urban grit. It is warm but not hot; there is an airy crispness (pleasant prelude to summer's unbearable humidity) and an unusual clarity of light.

I walk across Central Park, which even midday midweek is packed. Throngs of people run, ride, and roller-blade, and on the sprawling Sheep Meadow, where animals once grazed, a great herd of humanity has gathered and happily passes the time playing Frisbee, sunbathing, and dancing to the music of a boom box. Museums launch their biggest shows: Leonardo, Picasso, Matisse, and Delacroix; the opera season is winding down while the ballet season is picking up steam, and Broadway theaters are packed despite a musicians' strike. So much life and energy, so much to see and do; what an amazing city!

I go about my regular activities but view everything differently, with a new fondness and appreciation—precisely because I am leaving.

Toward the end of my last visit in Italy I ran my idea by Ivana.

"What would you think if I came here to live?"

"Come again?"

"What do you think about the idea of my coming to live *here*, like maybe for a year or something?"

She seemed a bit taken aback.

"Well, of course I would like it if you were here. But what about *you*? Wouldn't it be hard for you to leave your home and your family and your friends and come to live in a strange country? It is one thing to come and visit every now and then but maybe a very different thing to come to live, no?"

"Yes, I know. You may be right. It might be hard—it would certainly be different. Maybe I would hate it. But I have thought about it and am willing to give it a try—if it's okay with you, that is."

We quickly let the discussion drop, allowing the idea to settle in for both of us. But a seed had been planted. Before I left, I planted another.

"If I came here to live, would you live *with* me?"

This one was more of a bomb than a seed. While Ivana didn't totally rule out the possibility, it clearly made her uncomfortable.

"*Senti, amore mio.* If you were here, it would make me very happy. I like to be with you. I would gladly come to sleep with you, maybe even every night. But I couldn't live with you; I couldn't move out of my home, away from my mother, and move in with you, at least not right away."

At first I was disappointed, but then I thought about it. It might be a run-of-the-mill thing in the States, but in Italy two people living together—unmarried people—is still pretty uncommon, especially outside a city, especially in a small town or rural area, and *especially* when the woman is a native and the man a foreigner.

Moreover, I reasoned, Ivana was still feeling the sting of a bad marriage. She left her home in the Bussia to get married, spent all her savings on a fancy church wedding, and ended up nearly suffocating in an apartment in the Alba police barracks. She gave up everything for love and got burned. No wonder the idea of living with someone again, maybe even the possibility of falling in love again, scared the hell out of her.

And what about me? Was *I* really ready to jump back into a full-fledged relationship? Our connection was strong but it was still new. Wasn't it possible that the pressures of cohabitation might put too much of a burden on our still budding rapport? Things were going well. I was still being happily pulled along by that gravitational force I had felt months before on the road down into the Bussia, except that now it had been given a name: *siamo in due*. Why force it? *Piano, piano,*

I told myself. I thought it over. Then I regrouped and decided to go ahead with my plan.

I returned to New York, but this time it was different: I was on a mission. I emptied out my rented storage room, threw away mountains of old records from the restaurant, and sorted through tons of books, photos, personal memorabilia, a life's worth of love letters, journals, trinkets, and stones collected on the beach near my childhood home (I kept almost everything). Then I rented a mini-van and made two trips up to a friend's house in Connecticut who had offered to let me keep stuff in the attic.

Back at my apartment, I took pictures down from the walls and packed them carefully away. I cleaned places that had never been cleaned before and mercilessly went through all my belongings, determined to eliminate any unnecessary clutter.

At the same time, I spoke with real estate brokers to try to find someone to sublet my apartment. The situation looked grim. Sublets were way down, especially furnished ones. I had hoped to be back in Italy for Easter, but the clock was ticking and everything hinged on getting someone to rent my place. In the midst of total chaos, I went to Florida to visit my father, whom I hadn't seen in almost a year. Things there were not going too well; he was battling Alzheimer's and slowly but surely losing. While there I got a call on my cell phone from a real estate broker who had someone she wanted to show the apartment to. "Fine," I said, "get the keys from the doorman," and hung up. Later that same day I got a call back from the broker. The person loved the apartment and had agreed to the asking price.

I met him as soon as I got back, and he was perfect: friendly, polite, impeccably dressed, and thoroughly French. His company, a major Parisian fashion house, was sending him to New York for a year to launch a new product line. In no time, it was a done deal; he had a place to stay in New York, and I had rented my home for a year and was going back to Italy.

It is hard to pack up twenty years of personal belongings in two weeks, especially without having anywhere to unpack them to. But I did it.

I went to Good Friday services at Trinity Church, where the wonderful choir sang a motet by Francis Poulenc (*Vinea mea electa*) that seemed chosen just for me:

My favored vineyard, I planted thee . . . I fenced thee and removed thy stones and built a tower in thy defense.

And I went back again for Easter Mass on Sunday. After the starkness of Lent and Holy Week (during this period the altar is stripped, candles are extinguished, and the priests wear simple muslin robes), the church was ablaze with flowers and light. It was packed to overflowing, and everyone was decked out in their best Easter outfits. The service began with a joyous outpouring, a sun shower of alleluias, and I belted out the familiar festive hymns at the top of my voice: *Christ is Risen! Spring is here! I'm going to Italy! Alleluia!* Afterward, I floated home along the Hudson River path.

Three days later everything that needed to be done was, and then some: I even went to the unusual length of throwing a small going-away party for myself in the outdoor garden of a favorite restaurant. I took a deep breath; I played my piano one last time while I waited for the car to come and take me to the airport. And then I left.

After the frenzy of the past few weeks, a peculiar calm descended as I boarded the plane for Milan's Malpensa airport. I sat through the long flight in a tranquil daze. I was happy, at peace with the major change I had undertaken and the new experience I was embarking upon.

Fabrizio and Ivana were to meet me at the airport since this time I was not going to rent a car. I breezed through Immigration and Customs (surprising, considering the then-raging fear of the SARS epidemic) and waited for them outside.

When I first saw Ivana, there was that odd feeling of surprise and recognition you get after a long absence, the compound sensation of nostalgia, imagination, and actuality. Here, in the flesh, was the person I had made love to and dragged myself away from two months before, and here I was too, not just for another short visit but to live, at least for a while. We exchanged a major hug (I feared I would break her in two), then I loaded my huge suitcase into the trunk. She sat in the back. While the three of us made idle conversation, Ivana and I held hands through the seats.

Once we reached the Langhe, I was pleased to see that the irrepressible rites of spring were in full swing here as well. It was a glorious transformation: the bare, dusty brown- or snow-white-covered tongues of earth were now draped in soft velvety green; bright geraniums decked every balcony in Castiglione, and the cobblestoned lanes were scented with jasmine, lavender, and honeysuckle.

I had booked a room above the small trattoria on the main piazza for a month or so while I looked for more permanent lodgings, and that night Ivana and I celebrated our reunion with renewed vigor and intensity. As luck would have it, the restaurant was closed and we were the only overnight guests; the bedsprings squealed with joy while outside a pair of doves cooed softly and the song of a lone nightingale drifted in through the open window. I was back, and my long Lenten penance had finally come to an end. *Alleluia!*

I woke early the next morning, jet-lag free and ready to go; there was work to be done. I met Fabrizio in the vineyard and was amazed at the transformation that had taken place there as well. The alleys of mud and frost between the rows of vines had become soft carpets of grass. And the *gemme* on the brittle brown *tralci* (vine stocks) I had helped tie down were sending up tender green shoots about a foot tall.

It was a balmy day and Fabrizio and I were both wearing short-sleeved shirts—a far cry from the multiple layers of my past visits. Argo the German shepherd was there too and still sporting his heavy winter coat. But the "monster" that had terrorized me the night of my first visit to the Mascarello home turned out in the light of day to be a harmless, affectionate lamb. He wagged his tail with excitement when he saw me and lay at my feet as I worked. He knew me now.

The first of many spring chores is called *scarzolé* (not to be confused with *scalusé*, the initial righting of the wooden posts); it's the first round of green pruning that will be repeated several times during the coming months. Unlike the harsh clipping I did the first time I ventured out into the vineyard, this soft, gentle work is best accomplished with bare hands. You pick off the first couple leaves of each branch and pop off any little shoots that have emerged from the vine. The goal is to eliminate any stray growth—literally nip it in the bud—especially on the old part of the vine below the *capo a frutto. Scarzolé* helps focus mineral and energy resources up into the all-important grape-growing sections of the vine and prevents the formation of unnecessary branches that would eventually block light to the emerging grape clusters.

It's easy work; it doesn't require the level of concentration that pruning or tying the vines does, and Fabrizio and I chat freely.

"So, you are back."

"*Suna sì!*" I reply in Piemontese. "Yes, here I am."

"You like to be a *contadino*, eh?"

"Well yes, I really do enjoy working with you here in the vineyard. But I know being a farmer is a hard life. It might be very different if I had to do it every day, day in and day out, in order to survive. And everything hangs on the weather."

"Yes, hard it is. Take Maria, Bruna's mother; that woman has worked more hours in her long life than I ever hope to in mine, hard physical work. And even at her age she does it still, out in the vineyard at the crack of dawn and the last one back after dusk; she's like a machine that just doesn't stop.

"And they're shrewd too," he continues. "These old *contadini* might dress shabbily and live in run-down farmhouses, but many of them actually have loads of money stashed away in mattresses and in holes in the floor. Plus, their land is worth a fortune! There's a saying here; '*fare il contadino*,' that is to be *furbo*, sly as a fox. They might not say much, but they're tough as nails and always find a way to get what they want."

Then after a pause: "You don't miss the big city? I've never been to New York but I imagine it must be quite different from Castiglione Falletto."

"Yeah, it's about as different as you can get. But I like it here. It feels surprisingly comfortable and easy, the transition. I don't miss New York that much, at least not yet."

I tell him of my plans to apply for Italian citizenship, which I had learned I might be entitled to since my grandfather was born in Italy and had a son in America (my father) *before* he became naturalized.

"So you want to stay here forever?"

"Oh, I have no idea. But at least having an Italian passport would allow me to come and go as I please. 'Forever' is a long time. It all depends on what happens, just like how these grapes turn out depends on what happens over the next four months."

"Funny," Fabrizio continues after a pause. "For so long so many Italians have left home to find wealth and success in America.[1] We have another expression: 'trovare l'America.' When someone has had some sort of windfall or stroke of good luck we say, 'Ah, you have found America!' Now you, an American, want to come here and sweat in the vineyards. *Ha! Che strana la vita!*"

"Yes, life is strange, but it's all relative. Each person must find their own happiness in the time and place that is right for them. Who knows? Perhaps I'll find my *'America'* here in the hills of the Langhe."

12 · ROOTS

ONE OF THE most traumatic periods in the life of the vine took place in Europe during the late nineteenth and early twentieth centuries, and the culprit was nothing but a squashable little bug.

Phylloxera is a type of aphid. In late spring and summer it attacks the supple leaves and stems of burgeoning vines; when it has got just about everything worth getting up there, it starts to work its way down the trunk and, just as it's starting to get cold outside, into the earth. It tunnels through the soil and attaches itself to the roots, puncturing holes and sucking out the sap, the lifeblood of the vine. Afterward, when the vine is dead and rotting, the bug takes a nice long nap, wakes up in the spring, and crawls up into the light of day as if stretching in the warm sunshine. There it lays eggs and dies. The second generation, oddly enough, grows wings, and becomes airborne. When one area has been devastated, the aphids move on to the next—and the next and the next.

Like most insects, phylloxera reproduce rapidly and exponentially, especially when well fed. And here, in the vineyards of Europe, they found their Shangri-La.

Throughout the seventeenth, eighteenth, and nineteenth centuries, the European wine industry had continued to grow. The demographics of grape-growing, wine making, and wine consumption became ever more clearly defined as they continued to expand, while the crossing of geographic or political boundaries became more frequent and commonplace. Great Britain abandoned its own early attempts to grow grapes and instead became a major importer/consumer of French claret, Portuguese sailors (adding brandy to barrels of mediocre wine in an attempt to stabilize them for long sea journeys) inadvertently invented port, and the English, having developed a fondness for this fortified wine, jumped in to market it; French tastes and techniques of wine making crept more and more across the Italian border along with the French themselves, while, back on his Virginia plantation, founding father Thomas Jefferson planted *Vitis vinifera* from cuttings he himself had brought back from France. Many French winemakers looked with great interest upon the New World's hearty wild American vines and imported cuttings to plant and experiment with. This innocent activity was to have catastrophic consequences for the European grapevine.

By the mid-1800s, wine had become an important international economic factor; production had reached new highs, and there was more land under vine than ever before. So when phylloxera first turned up in France in the vineyards of the southern Rhone in the 1860s, some producers took it very seriously. Others, in denial perhaps, simply ignored it in the hopes it would just go away. But it didn't go away. More and more attention was given to the rapidly developing crisis, rewards were offered for a successful way to combat the disease, and numerous treatments were tried. But nothing worked. *Vitis vinifera* was defenseless, and the ravenous bug appeared to be unstoppable.

Over time, phylloxera spread throughout Europe, destroying more than 50 percent of its vineyards and posing a serious threat to viticulture throughout the world. The situation was looking grim indeed. Finally, just in the nick of time, a cure was found, and it came from a most unlikely place. It was determined that indigenous wild American vines (*not* members of the noble *Vitis vinifera* family) were resistant to phylloxera and that European vines grafted onto the roots of American ones were of no interest to the greedy aphids.

Grafting is an old technique, a kind of hybridizing, used in many different types of agriculture, and there are several ways to do it. Basically, a notch is made across the main stem of the parent vine just above the root, while the cut end of the branch to be grafted (known as the scion) is sharpened to a wedge-shaped tip. The wedge is inserted into the notch and securely tied into place. If correctly done, within a short time the cells in the cut ends of both vines fuse together into one plant with qualities of the previously separate two, in this case the phylloxera-resistant roots of the American vine and the great *Vitis vinifera* grapes of the European one. (The same technique can also be used to change the grape varietal on an existing vine without having to rip the whole thing out and start all over from scratch.)

Despite initial resistance (Would American roots change the nature of European grapes? Could a New World upstart really be the savior of the Old World viticultural tradition?), grafting vines onto American rootstock was soon found to be an effective way of stemming the catastrophe and was quickly adopted by European growers. In Bordeaux, by the end of the nineteenth century just about all of the old prephylloxera vines had been ripped out, grafted onto American roots, and replanted. (The necessity of replanting also created a good excuse for the implementation of other new and improved viticultural techniques on a widespread basis.) The bug turned up in the Barolo area in the early decades of the twentieth century, where the events followed the same pattern but in a much shorter span of time.

For many it was too late: some growers were completely wiped out, while others who couldn't afford to purchase resistant roots and replant their entire vineyards had to cut losses and sell out. Phylloxera had taken its toll. Nevertheless, the introduction of American rootstock proved to be the silver bullet that kept the voracious insect in check and saved European viticulture.

Sometimes as I stand in the vineyard and look around at the endless rows of thriving vines, I remember that these vines, and the shoots and grapes that will emerge from them, are sprouting up out of the Italian earth from American roots. This makes me, an American, feel proud (I quickly dispel the thought that phylloxera probably came from the United States in the first place); it gives me a sense of a broader, deep-rooted historical connection to this place and to this activity of grape growing. It also makes me feel like I am reversing the process just by being here: in a way, I am attempting to graft myself onto Italian roots and adapt myself to this new environment. Who knows whether it will take hold and bear fruit as successfully as the widespread grafting, which took place here nearly a century ago?

13 · BETTER LATE THAN NEVER

I NEVER KNEW my paternal grandfather (or any of my other grandparents for that matter) as he died shortly after I was born. But, having learned that being his descendant could be my best ticket for an extended stay in Italy, I began accumulating the documents necessary to apply for Italian citizenship.[1] And as these documents began slowly drifting in, I got to know him a little.

His name was Vincenzo, which comes from the verb *vincere* and means "the winner" (combined with his last name it means the "late winner" but, as the saying goes, *Meglio tardi che mai*).

In the late 1800s, just around the time phylloxera was starting to show its evil face in Europe, Vincenzo was born in a village called Acerra in the province of Naples. However, the exact date of his birth remains a mystery. The document I received months after making the original request for his birth certificate was not an actual certificate but rather an *estratto,* an official extract of information that had been entered by hand with a quill pen and blotchy ink into the thick, dog-eared register of births and deaths in the town hall well over a century ago. The date on this *estratto* was different from that on an older Xerox copy of one my dad had obtained twenty years earlier when he visited Acerra, and it conflicts with every one of the other various dates of birth listed on all the other documents of my grandfather's life.

There are a number of possible explanations for this confusion: in extracting information from an old book, it is often difficult to accurately make out all the dates and spellings, plus there is always the possibility that someone can make an innocent error in the process. Moreover, this being southern Italy and Acerra a particularly infamous stronghold of organized crime (known here as the Camorra), there are often other reasons for less than rigorous accuracy regarding dates and other personal information.

Apparently, there are many, many Tardis in Acerra, and a number of those born around this time were named Vincenzo.

Vincenzo's father's name was Raffaele, which was also the name given to his oldest son (my uncle Ralph, a k a Fuzzy). It's my middle name too. Until I got the *estratto*, I never knew the name came from my great-grandfather.

Early in 1903, Vincenzo left Acerra for the United States. The reason for his departure was unclear but it was probably the obvious one; Italy, especially this southern region, was dirt poor and many left in hopes of finding their fortunes—or at least a fighting chance to survive—in America.

He arrived in New York harbor on February 14, 1903 (exactly one hundred years before Ivana's mother would have her heart operation), and after being processed through Ellis Island with the multitudes of other tired, hungry, and oppressed immigrants from all over the world, made his way to Chicago. There, Vincenzo settled in the Italian ghetto on the South Side along with other transposed Italians, many of whom also happened to be from Acerra.

This immigrant community was like a small southern Italian village superimposed onto the ramshackle tenements of the Windy City. Laundry hung from fire escapes, the aroma of bubbling tomato sauce wafted through paper-thin walls and *pizzaioli* plied the crowded, dirty streets, singing passionate arias in Neapolitan dialect to advertise the warm pies stacked up on their heads. Here the putrid smells of the nearby stockyards and dead fish washing up along the edge of Lake

Michigan replaced the olive-scented Mediterranean breeze, and gangsters used pistols and automobiles instead of knives and horses.

But life was still better here than in the old country. Here there was work; here at least you could eat, and as long as you minded your own business, the tough guys mostly left you alone. They left Vincenzo alone anyway; he was from Acerra and he *knew* people.

Vincenzo went forward with his new life in America and never looked back (though he surely dreamed every now and then of running barefoot through the sun-washed hills covered with low craggy olive trees or rolling in the dusty vineyards). He never learned to speak English—in fact, he never really spoke Italian either, just his Neapolitan dialect—and he never learned to write except to sign his name. But that was okay; everyone in the neighborhood understood him, and he didn't talk that much anyway.

According to the documents, Vincenzo was five feet six inches tall (like me) and had brown eyes and black hair. He was an unskilled laborer, a manual worker, and, along with throngs of other immigrants—Italians, Poles, Irish, Germans—helped dig the subway tunnels of Chicago.

In 1914 he married a woman named Carmela Calabria from his hometown (it is not clear whether they had known each other back in Acerra, but it would seem likely), and they had children, five in all, of which my father, born in 1919, was the second. His parents' dialect was my father's first language, but he quickly learned English once he started going to school and became the de facto translator for his mother and father. This responsibility usually fell to him as his older brother (my namesake) was often out on the streets or in trouble with the law. (Many years later, when my dad was working as a psychiatric counselor in the Illinois State Penitentiary, an appreciative inmate asked him if he would like my uncle's prison record to "disappear"; he thanked him but declined.)

Vincenzo become a naturalized American citizen on February 25, 1925, when my dad was five years old. In the naturalization documents he renounced his allegiance to Victor Emmanuel III, King of Italy, and affirmed that:

"I am not an anarchist; I am not a polygamist nor a believer in the practice of polygamy; and it is my intention in good faith to become a citizen of the United States of America and to permanently reside therein: SO HELP ME GOD."

These affirmations were explained to Vincenzo by one of his two elder children, who were there with him that day. Then without further ado he signed the document in his own shaky hand and, turning his back once and for all on the country where he was born, raised, and where most of his family still lived, officially became an American.

I doubt he ever regretted it.

Becoming naturalized was the "natural" thing to do, after all. His life was in America now: he had yanked himself up by the roots and transplanted himself to the concrete and asphalt of the South Side; there was no going back. Besides, by becoming a citizen he was just catching up with his own children, who, having been born on American soil, were already Americans; *they* didn't have to renounce anything to get it. (In addition to being Americans by birth, Vincenzo's children, born before he became an American, were also automatically Italian citizens by blood, though I don't think any of them ever knew it or would have cared much if they had.)

The one and only photo I ever saw of my grandfather was taken in Chicago, probably in the 1930s. This grainy image shows him leaning up against a brick wall in the bright glare of sunlight wearing high-waisted, wide-pleated slacks, a starchy white button-down shirt with the sleeves rolled up above his elbows, and a wide-brimmed fedora hat, which cast a shadow over his eyes. He had a thick salt-and-pepper mustache that was slightly angled upward on one side by the barest hint of a smile or smirk. He held a wine bottle in one hand and a lighted cigarette in the other.

After he came to America, my grandfather never returned to Italy; he never left the United States and, as far as I know, never even left Chicago—in fact, he rarely left the neighborhood. The graft, evidently, had taken. Vincenzo, the "belated winner," died in his adopted home in 1959. He was sixty-eight years old, or seventy-seven or seventy-nine or eighty-eight, according to the various documents: no one was quite sure exactly when he was born or how old he was when he died—and that was just fine with him.

14 · BEATING HEART

BACK IN THE Langhe, spring is here, and sleepy Castiglione, like the buds on the previously sleeping vines all around it, has sprung to life.

The room I have rented, one of four above the *trattoria* Gran Duca, is in an old building on the Piazza del Centro. While the name may seem rather grand for what is really just an open intersection with space to park a handful of cars, it is perfectly appropriate. This *is* the center of the village, both geographically and psychologically, the place where most of its businesses are located (four plus the nearby Bancomat) and where much of the activity takes place. In fact, this unimposing patch of asphalt constitutes the beating heart of this little community.

Judging by the way they gently curve around the castle periphery like the spiral of a seashell, the Gran Duca and the buildings on either side of it probably mark the original limits of the ancient fortified town, and the empty space in front (now the piazza) was probably once the site of a drawbridge or protective gate. My room looks right out onto the piazza, just across from where the main street, Via Roma, comes up from the Alba-Monforte road and takes a sharp turn to the left as it heads up to and around the castle, before turning back into itself to form a giant *P*. (Castiglione Falletto is a dead-end town, and, except for paths through the vineyards, the only way to go out is the way you came in.)

The Gran Duca is owned by a young guy from Sicily with oily black hair and smudged glasses who moves stealthily about, while a filtered cigarette wedged between his thin lips indiscriminately drops a trail of ashes. In the morning, we often have espresso together in the little breakfast room downstairs. In the evening, he works the dining room (when he feels like working, that is) and can even be found some afternoons preparing something in the kitchen. He is assisted in the restaurant by two women, both of whom are also transplanted southerners. Lucia cooks and does a decent job at it, though she misses the zesty hot peppers and ubiquitous sweet-sour tomatoes of her birthplace, while Anna washes the dishes. Anna is shaped like a diminutive bowling pin and usually wears a festive flower-print dress under a frilly apron. She always has a twinkle in her eye and a genuinely innocent smile on her face, which endears her to just about everyone.

You have to walk right by the open kitchen door and through the dining room to get upstairs to the guest rooms and, unless it's really late and everyone has already left, Ivana and I say good night to everyone, employees and diners alike, on our way up to bed.

Next door to the Gran Duca is the *tabaccheria,* though the label is deceptive for tobacco is just one of many items for sale. Besides cigarettes, cigars, lighters, matches, and other assorted smoking paraphernalia, the store stocks newspapers and magazines, books, gifts, dolls and toys, toiletries, pajamas, umbrellas, kitchen utensils, souvenirs, postage stamps, stationery, a selection of blurry but colorful postcards of Castiglione, which appear to have been taken many years ago, and bolts of fabric. (These are just some of the items on display—who knows what else might be stashed away in back!)

The *tabaccaia* is short and bulky with big features and mussed hair. She typically wears a long skirt or overcoat with a scarf wrapped around her neck, giving the impression she is wearing a robe. Outside of her shop she doesn't socialize much with the other villagers, some of whom think she is a bit odd. She has always been very nice to me, though. I like to chat with her when I go in to buy my newspaper. Although she knows me well by now and always greets me with a dour smile, she refuses to address me with anything other than the formal polite *Lei,* despite my repeated requests for *tu.*

The *tabaccaia* is almost always accompanied by her thirty-something son; if they are not together in the cramped store, they are sitting side by side on the bench right outside it. I have tried to chat with the son too, but on the rare occasions when I have encountered him alone on his way to or from the shop and said hello, he remained stone-faced and unresponsive as if he hadn't heard a word. When he is not with his mother, he is often to be found sitting in his car in front of the shop, sometimes for hours on end, with the radio blaring. Sitting on the dashboard is a little teddy bear whose ever-smiling face gazes adoringly upon the son seated behind the wheel, much as the son gazes adoringly upon his mother.

The husband, *il tabaccaio,* is never in the store at the same time the two of them are. He is more talkative than his son but less so than his wife. I have chatted with him on occasion and even bumped into him in the vineyards (he doesn't have land of his own anymore but sometimes helps others). He always has a wet, stubby cigar stuck into his mouth, and his few remaining teeth are either tar-blackened or gold.

On the other side of the piazza, next to the house where Bruna was born, is the *panetteria,* the bread bakery. This label is a bit misleading as they don't actually bake anything here anymore (an assortment of bread is delivered five days a week from a commercial bakery). Besides bread, the store offers a wide range of staples from soap to salami; a variety of cold cuts and cheeses, sausage (occasionally), a small selection of fresh fruits and vegetables (plus a freezer full of frozen ones), not to mention nonperishables like garlic, potatoes, and onions, canned goods, dairy products, salt cod, household cleaning supplies, mineral water, beer, and wine. They sell loose eggs, which they wrap up in newspaper for customers to take home, and they'll make a sandwich for you if you want one.

This business, like the tobacco shop, is also operated by a trio of husband, wife, and their unmarried adult son. The son, who likes to have the sides of his head shaved to eliminate sideburns (making him look somewhat like a monk in street clothes), showed no interest in carrying on his father's craft and is often to be found around the corner in Renza's bar playing cards. The father dresses in a thick flannel shirt and wears a felt beret pulled down tight on his head. Though steadfastly retired, he still rises each day before dawn out of habit. It is said that he once baked excellent, hand-rolled *grissini* (breadsticks) in an oven in the back of the shop, attracting customers to his corner of the piazza from far and wide. But he got old, and, with no one to pass his expertise on to, the oven was removed years ago. It is rare now to see him in the store at all, as if he couldn't bring himself to sell the airy, tasteless bread and packaged breadsticks.

I asked him once if he ever missed baking:

"Ehhhhhhh," he exhaled in a long, trailing decrescendo, with a grimace on his face as if recalling his many hours among sticky doughs and hot oven, ending abruptly with a decisive "NO! I believe I made enough bread in my day to last a lifetime."

It would seem, however, that he does miss inhaling clouds of fine flour dust, for these days you can often find him in a chair on the stoop outside the shop smoking unfiltered cigarettes, which he buys three packs at a time from his friend across the piazza.

With father and son generally out of commission, it mostly falls to the wife to handle the daily operation of the business.

The shop may appear to be closed, as the lights are often off and the venetian blinds in the windows drawn. But don't let that throw you. If you take a chance and press on the door—and press you will sooner or later, for this is the only place in town to buy food—you will likely find it unlocked.

The opening door causes a buzzer to ring in back where the proprietors live and eventually brings one of them—usually the wife—out through a curtain into the store. She might appear somewhat disheveled and confused, as if caught by surprise that you are there. In the winter, her eyes are often red and watery, her nose stuffed, and you can hear the scratchy sound of rough skin as she rubs her hands together trying to ward off the cold.

The *panettiera*, like almost everyone else around here, seems to have considerable difficulty dealing with the euro: once you have finished selecting your items, she carefully tallies up the bill, calculating the total in *vecchie lire,* then converting it to the "new" currency and slowly counting out change to the *centesimo* with an uncertain look on her face.

The shop closes between one and three for lunch and at seven in the evening, as well as Thursday afternoon and all day Monday. But if you should ever need anything, you can always ring the bell and one of the proprietors will let you in through the side door to make your purchases.

*　　*　　*

From my perch above the piazza I watch a flurry of activity unfold below as if on a living stage. The villagers are a real-life cast of characters, and their entrances and exits seem almost choreographed: Franco, the athletic retired guy from Alba, parades around in his skintight yellow cycling outfit with a red bandana tied across his shiny bald head, singing loudly to himself; an old crippled resident with a crutch hobbles slowly but surely back and forth across the piazza, unfailingly stopping to salute me with a loud *"Oh! Cavaliere! Come va?"* whenever he sees me looking down from my window; the *panettiera* emerges from her shop early in the morning with a watering can and spins like a dervish to dampen the threshold to help keep the dust down; and Beatrice—designer-jeaned, redhaired, tattooed, and pierced youngest member of a trio of attractive sisters popularly known as the Three Graces—waves as she goes zipping by on her Vespa en route to some rendezvous.

Off to one side of the square a Greek chorus of old pensioners line up like pigeons on a bench under the balcony of an unoccupied house owned by Bruna's family and sit there all day watching and commenting on everything and everyone that goes by. This bench, made from a huge old varnished block of wood, is known as *Radio Trave* (*trave* means beam) due to the amazing rapidity with which news is broadcast from it. It has, I learned, another name as well:

The one (and so far only) time I sat down among the radio regulars, Franco leaned over to me and whispered:

"You know what they call this bench?"

"Sure. *Radio Trave*," I responded.

"Yeah, well, they call it something else too, *riunione dei molli*, the meeting place of the soft ones." You know what I mean? *Gli im-po-ten-ti*," he stressed for emphasis; "the impotent; those who can't do it anymore."

"Well, I hope it's not a requirement to sit here for a moment," I whispered back. "And I certainly hope it's not contagious!"

I wake up in the morning to the sound of tractors buzzing back and forth across the square on their way to the vineyards. And in the evening, everyone drifts around the corner into Renza's bar.

"Son' tutti pazzi—they're all crazy!" Renza says of her village neighbors, though if pressed she would have to admit she is one of them.

15 · RENZA'S

THOUGH TECHNICALLY NOT a Castiglionese (she was born in La Morra and lives in the hamlet of Annunziata across the valley), Renza is a sort of village matriarch and her establishment an unofficial but indispensable municipal institution. She is short and stocky, with pudgy red cheeks and wavy auburn hair, which she has washed, trimmed, and conservatively highlighted once a month or so at a salon in Alba.

Dressed in a nondescript below-the-knee skirt, dark becrumbed sweater, and slipperlike shoes,[1] Renza putters around the place speaking almost constantly in a singsong voice and almost always in Piemontese, regardless of whom she might be talking to. Sometimes, when she gets worked up over something or is waiting impatiently for a response from someone who doesn't understand her, she raises her voice, accelerates her speech, or throws up her hands for emphasis. Usually, though, she just shrugs her shoulders and breaks into laughter, turning her cheeks even redder.

She loves to chat and loves to offer hungry guests something tasty she has prepared on her little electric hotplate. She often praises her own culinary prowess—"No one makes a Pollo al Babi like mine!"—just as she often berates the card players for sitting there all night not drinking anything, the *ladri* in Rome for unbearably high taxes and inefficient government, and the *disgraziati* abroad for everything else that is not right with the world.

Renza is the repository of much local information; just about everyone stops by at some point or other, and she knows just about everything that goes on around here, both in public and behind closed doors. She used to have a small restaurant in Gallo, the commercial strip-of-a-town at the bottom of the hill on the way to Alba. But she has had the bar here for almost fifteen years, ever since her husband passed away. It is called La Terrazza, but everyone refers to the place simply as Renza's.

On the far side of the building is a lovely terrace that juts out above the vineyards and looks over the steeply sloped valley with the hamlet of Perno to the right and the tall, thin castle of Serralunga straight ahead. In warm weather, people come here to have lunch or a cool drink during the day or an ice cream in the evening. Tourists, especially German ones, like to sit here drinking wine all afternoon, chatting boisterously and taking photos of the bucolic countryside, while English hikers, sweaty and bright red from their final ascent up the footpath through the vineyards, stop to catch their breath in the shade of an umbrella as they nurse a warm beer or a cappuccino. When they get back home, many send Renza postcards or photos or scribbled notes, which she proudly sticks up on the wall inside.

Renza's interior is a long narrow rectangle. Just inside the door is a bar; on one end is the espresso machine and on the other a display case that contains an odd jumble of stuff available for purchase: Ferrero chocolates, candy, Pringles potato chips, parcooked cotechino sausage and lentils in a submersible pouch, T-shirts, and wine guides. The area behind the bar is in perennial disarray—dishes piled up in the sink, leftovers from lunch, a leg of prosciutto on the slicer—and above it all towers a wall of shelves holding mismatched glasses and an assortment of *amari,* grappa, and wine.

Along the opposite wall is a row of tables; the one right in front of the window is Renza's, where she often sits in her reclining plastic lawn chair during afternoon lulls leafing through magazines, doing crossword puzzles, and keeping close watch on comings and goings in the piazza. At the end of the bar is a tall refrigerator with sliding glass doors from which regulars help themselves to beer and soda, and beyond that the place opens into a comfy little room with four or five more tables covered in green-checked vinyl tablecloths.

At the rear of this room is a large TV that is almost always on. Often it's tuned in to the obligatory *telegiornale* news program or soccer match. But Renza prefers quiz shows, especially *Chi Vuole Essere un Milionario?* (Who Wants to Be a Millionaire?), which comes on at seven P.M., just before the after-supper rush, when she herself is finishing up her solitary dinner. She becomes totally engrossed in the game (woe unto him who desires a piddling little espresso when the fortune of a lifetime is at stake!), blurting out answers from the edge of her seat as if *she* were the one sweating under the bright lights.

"Perhaps," she says, sighing after the game is over, "one day I will go on this show. You saw it yourself, I got almost all the answers right, no?"

"Yes, you did!" I reply.

At night in her modest home across the valley, Renza dreams of being a millionaire, of dressing in fancy clothes, traveling around the world, and having others wait on her for a change. In her dream, a flattened chicken is presented on a fine china platter by a tuxedoed waiter while a piano tinkles softly in the candlelit background. She lifts a heavy silver forkful to her mouth, nibbles delicately, then slightly crinkles up her nose and muses: "Oh, it's not bad, to be sure. But no one makes a Pollo al Babi like I *used* to make!"

Then she wakes up, the dream fades, and she goes back to catering to the simple needs of her Castiglionese clientele.

Renza is as endearing as she is eccentric; underneath her buoyant, hard-as-nails exterior she is also fragile and sensitive. She can easily be made to feel undervalued and unappreciated—but not by me. Occasionally I have lunch there and praise her cooking. Sometimes I stop by in the afternoons just to hang out with her; I like to get her to tell stories, fill me in on local gossip, and teach me Piemontese. I stop by in the evenings too, when the place becomes a kind of communal living room; it is not uncommon for people to wander in after dinner in their *pantofole*.

Most evenings the tables in back are occupied by cardplayers. There is the usual cast of characters, and they break up into their customary groups to play either *scala quaranta* or *tarocchi*. I've learned the basics of *scala*, but *tarocchi* remains mostly a mystery. There are four suits— *spade* (swords), *denari* (coins), *bastoni* (clubs), and *coppe* (cups)—plus the *tarocchi*, elaborate depictions of personified figures such as the Angel of Death and Queen of Love.

The cards themselves are interesting to look at, and I like to watch the guys play even though I have little idea how the game actually works (Does Death take Love, or is it the other way around?). Once I actually got some guys to "teach" me while waiting for their fourth to show up, but it didn't help much: the rules are complicated to begin with and there seem to be endless exceptions and contingencies. Even the basic mechanics are challenging as you must hold twenty-two cards (twenty-five if only three people are playing) in one hand, something that everyone manages to do with precision and ease in their big rough bearlike paws, artfully arranging the cards as they are dealt and opening them up into a tight, wide fan.

While they play, people exchange news, chat, recite jokes, and tell stories. Occasionally there's a dispute; voices are raised and tempers flare, but it quickly dissipates in laughter. Sometimes money changes hands but only small change. There was a time, however, when the stakes were often much higher.

Many years ago, Maria, Bruna's mother, operated a tavern in their house on the piazza, where Bruna was born. "Oh, back then people would play cards all night long, sometimes for days on end. 'Bring us another bottle of wine,' they'd say, and I'd bring it, I would, and make sure I got paid for it too and right away while they still could. They'd play and play, hoping their luck would turn around, and lose and lose until they had nothing left. Sometimes they would wager the very clothes on their backs, their homes, their daughters. Even their vineyards! In the old days, fortunes were often won and lost over a hand of cards. *Erano proprio matti!*" she chuckled. (Maria knows what she's talking about; her husband originally won the place in a game of cards!)

These days, however, the former tavern is nothing but a seldom-used second home and over at Renza's people mostly play for coffee.

It was at Renza's one evening after supper that I heard about the oven.

Giuliano is a Renza-regular. He's a big guy with white hair, booming baritone voice, and hulking upper body tapering down into relatively skinny little legs. He's a farmer, cardplayer, former champion of the local sport, *palla a pugno,* and member of the church choir. He is also a lifelong resident of Castiglione and the town's perpetual vice mayor. ("One time he almost didn't get elected and was so upset he practically cried," Renza explained. "Now everyone automatically votes for him because nobody can bear the idea of that big guy bawling like a baby!")

As he nursed a *digestivo* waiting for his cohorts-in-cards to arrive, I mentioned there seemed to be quite a bit of activity in the piazza now that spring was here:

"Oh, if you think it is busy now, you should have been here forty or fifty years ago. There were lots more people around here back in those days. All the houses in town were occupied (now many people have moved away to Alba or Torino and use them only on the weekends, but back then they were all actually lived in). This bar here was packed all the time, and so was that trattoria where you sleep. At Maria's, people sang and danced until the sun came up. People stood around out in the piazza all day and went calling on each other in the evenings.

"There was lots of activity back then," he recalled, "and no cell phones or televisions, so you *had* to talk and visit. Cars were still a novelty. People came here from all around to shop and socialize, many on foot or bicycle or horseback; plus there was still the *furn*."

"The what?" I interrupted.

"The *furn—forno* in Italian—the village bread oven that used to be right there on the piazza."

"You mean the one in the *panetteria* shop?"

"No, that was just a small oven for making breadsticks. I mean the *big* village bread oven. Almost every commune had one. People used to come here from all over, even from Monforte, La Morra, Grinzane, and Serralunga, to get bread, and sometimes they brought stuff too— hunks of meat or animals they shot or casseroles—to slowly cook in there as the oven was cooling down. I can still remember waking up to that smell of wood smoke and baking bread every day, six days a week. And the taste . . ."

"What happened?" I asked.

"Oh, the baker died," someone else piped in. "He was an old guy. His wife had passed away years before, and his kids moved away. In any case, he lived by himself. He just got too old to wake up at four in the morning to light the fire every day and mix the dough.

"One day, on a Tuesday right after Christmas, we woke up and there was no smell of fire (you get used to that smell, you know, just like the sound of the campanile ringing all night long; it gives you a pleasant sense of security and warmth, even in the summertime, and you miss it if it's not there). Well, there was no smell, so people went over to his house and found him there, dead, seated at his little table with a bowl of cold soup and a piece of stale bread in front of him; he even had a spoon in his hand like he was just getting ready to eat. He had passed on at some point after Midnight Mass; that's when people last remembered seeing him. They found him there two days later. That's why they still call that little run-down house where he lived the Casa della Morte."[2] I knew the house he referred to, up behind the castle; it is untouched and unoccupied to this day.

"We had a simple funeral for him just before New Year's. And do you know what? He had made his own coffin right out of bread! He had baked long flat thick panels of bread (probably from leftover dough) in that big old oven, left them to dry, and then stuck them together. We put him in there like he wanted and carried him all the way from the church to the cemetery. He didn't weigh very much anyway. Ha! What do you think of that, a baker buried in a big loaf of bread! I've never seen anything like it! But then he was Castiglionese after all."

This guy was on a roll, so to speak; there was a twinkle in his eye, and he occasionally pounded the table for emphasis. I don't know how much of what he said about the old baker was true—people like to tell stories, especially to a gullible foreigner, and it's not always easy to tell where fact trails off and fiction begins. But then again people from this town do have a reputation for being a bit peculiar, not to mention practical and thrifty.

"After that," Giuliano continued, "no one ever took his place. They didn't need to. Almost everyone had cars by then, and you could have bread delivered or go get it from the supermarket in Gallo or Alba. Still, I kind of miss that smell, come to think about it."

"What about the oven?"

"Oh, it's still there, as far as I know. You know that little one-story building on the piazza right next to the *tabaccheria* where the road turns? Well, that's where it was, right inside those large wooden double-doors—probably still is; I don't recall anyone ever taking it out."

"Hmm, I'd like to have a look inside that building one day," I said aloud, but he didn't hear me: the others had arrived and seated themselves at the table, and Giuliano was busy dealing out the first hand.

· RECIPE #8 ·

RENZA'S POLLO AL BABI

Renza's Chicken, "Frog Style"

HERE, AFTER MUCH cajoling, is Renza's recipe for Pollo al Babi. (*Babi* is the Piemontese word for *rospo*, or toad, so named because the flattened chicken resembles a frog.) She would like to make it perfectly clear, however, that her recipe is intended solely for the use of private individuals or families outside Italy and that its use ("theft" was her word) by other chefs or restaurateurs is strictly prohibited. She would also like to warn unscrupulous professionals that she has anonymous inspectors throughout the world who are constantly on the lookout for violations and that any infractions will be vigorously prosecuted.

Furthermore, Renza would like it duly noted that she offers the following recipe only out of generosity and goodness of spirit (and perhaps to keep a certain American from pestering her). She makes no guarantee, however, that it will work (she herself doesn't use recipes) or that what you make will be as good as hers—in fact, she rather doubts it. In any event, here (with a few adjustments) it is.

INGREDIENTS

1 large chicken (preferably fresh, free-range), about 2½ pounds
½ teaspoon dried rosemary
½ teaspoon dried oregano
½ teaspoon dried thyme
1 dried bay leaf
1 pinch of hot pepper flakes (optional)
1 pinch of garlic powder
1 pinch of cumin powder
Fresh black pepper to taste
1 pinch of coarse salt

2 tablespoons extra-virgin olive oil
3 thick slices pancetta or smoked bacon, cut crosswise, into ¼-inch strips
2 tablespoons white cooking wine
Juice of one lemon
Grated zest of one lemon
1 tablespoon coarsely chopped Italian parsley
Fresh rosemary and lemon wedges to garnish

1. Using a sharp chef's knife, split the chicken down the back on either side of the neck and remove the backbone (discard the bone or reserve for stock). Remove the sternum between the two breasts, open the chicken, and flatten it out. Make two small slits in the skin at the bottom of the chicken on either side of the backbone and insert the tip of the drumsticks into each slit. Tuck the wing tips inside the wing. (You may ask your butcher to do this for you.) The chicken will resemble a frog.

2. Put the rosemary, oregano, thyme, bay leaf, and pepper flakes (if using) in a spice grinder and grind to a fine, even consistency. Remove any large pieces and add the garlic and cumin powders and black pepper.

3. Season both sides of the chicken with salt. Brush or rub the skin side of the chicken with a little bit of the olive oil, then sprinkle liberally with the herb-spice mixture.

4. Put the pancetta in a large heavy-duty skillet over a medium flame and cook, stirring occasionally, until it begins to color.

5. With a slotted spoon, remove the pancetta and set aside, leaving the rendered fat in the pan. Add as much of the remaining olive oil as necessary to cover the bottom and place the chicken skin side down into the pan, and cook, moving occasionally to make sure it doesn't burn or stick, until the skin begins to turn crispy and golden brown (approximately 20 minutes).

6. Carefully turn the chicken over. Place a cover over the pan—making sure, however, that it is askew so air can get out (otherwise it will generate condensation and the skin will lose its crispness)—and continue cooking for approximately 20 to 30 minutes, or until a knife goes in easily and the juices that come out are clear.

7. Transfer the chicken to a serving platter and keep warm. Carefully remove any excess fat (there should be about one tablespoon of fat plus some juice left in the pan) and raise the heat. Put the cooked pancetta back in the pan, then deglaze with the white wine, scraping the bottom of the pan with a wooden spoon to remove any caramelized bits. Add the lemon juice and, if necessary, a touch of water.

8. Drizzle pan juices over the chicken, sprinkle with the lemon zest and chopped parsley, and garnish with fresh rosemary and lemon wedges.

YIELD: SERVES 2 TO 4 AS A MAIN COURSE.

16 · IRREPRESSIBLE URGES

MANY OF THE guys I see hanging out in Renza's bar in the evening are the same guys I see driving tractors back and forth during the day, caps pulled down over their eyes to shield them from the blinding sun.

This is a period of intense activity in the vineyards and, except for a midsummer lull, it will not let up until fall when the harvest is over. The principal tasks of winter, pruning and tying down the vines, are critical but intermittent: you can do them pretty much when you want (the pruning anytime between the end of the harvest and February, the tying anytime between the pruning and the end of March, so long as it's before the buds start to bloom), and once they're done they're done.

Throughout the spring, however, there are a multitude of chores and the vines require—*demand*—almost constant attention. Once the first green pruning *(scarzolé)* is finished, a pair of nylon cords is run from post to post along the length of the rows just above the first set of metal wires. The new shoots are gently tucked in between the two cords to help train them upward; this is called *tirarli su* (to lift them up). Weeds and other stray growth must be removed to prevent them from depleting the vines' valuable resources.

At this point the vines are like newborn babies, fragile and vulnerable, but amazingly vigorous, bursting with energy. They soak up everything around them—water, soil, sun—and seem to grow right before your eyes. During this time the farmer is like a conscientious parent, coddling, nurturing, encouraging, guiding, fussing over; trying as best he or she can to take care of the tender shoots and infant grapes, paving the way for them to grow strong and healthy and realize their full potential. Then, before you know it, the helpless babies turn into precocious, promiscuous adolescents.

No sooner is this first pass finished than it's time to start all over again: a second round of *scarzolé,* another set of nylon cords just above the second wire, more weeding. Once again the slender branches, now nearly two feet tall, must be carefully threaded between the new set of cords with minimal disturbance to the tight little bunches of nascent green grapes to be, which have only just erupted into delicate flower.

The vine, like humans, is a sexual organism but (unlike us) a hermaphroditic one; it pollinates itself with its own flowers, the tiny grains of pollen moving from the blossom's male portion (called the anther) to the ovaries of the female (via the opening called a stigma) aided by the wind, a variety of insects, and the rustlings of man.

With the first pruning, you cut and pull and discard; then you gently but firmly bend, cajole, and tie down in place. Now, however, you gently *caress* the flowering branches.

I pause from my work in the vineyard to straighten up and take a look around. The lush landscape looks peaceful and placid on the surface, but, in actuality, the same thing that is going on with the vines is going on all over the place: an irresistible urge, a mounting frenzy of desire, a rampant orgy of sex and reproduction. Clouds of pollen fly wildly through the fragrant air indiscriminately searching for something to impregnate. Flowers burst into bloom, releasing attractive scents and blatantly exposing their floral genitalia, while bees buzz frantically from one to the next as if drugged, and weird insects lounge languorously under supple green leaves, which have graciously unfolded themselves to the warm caress of the sun. Back in Castiglione, village mutts sniff the air, lick foaming chops, and strain at leashes while members of the extended family of inbred black-and-white cats slink around, purring with half-closed eyes and odd grins on their whiskered faces.

Ivana and I too succumb to the persuasive impulses of the season throbbing all around us (though we both are careful to avoid the reproductive part). After a month in my perch above the piazza, I move back to the hotel while I continue looking for a suitable long-term rental, though this time I get one of the large apartments with a kitchen. And Ivana follows me there. We have spent nearly every night together in a variety of different places; practically every room of the hotel, two different rooms above the piazza, and now this apartment.

I now take her coming to me each evening as a given, though I never take it for granted. The sense of newness and mystery has faded some-what, but the intensity of our lovemaking has not. On the contrary, like a musician who can only really begin to play once fingers have learned the notes or a frequent traveler who can only begin enjoying the partic-ularities of the landscape once the curves and dips in the road have be-come second nature, it has increased. As we have gotten to know one another better, our feelings have blossomed and taken flight like the puffy yellow flowers on the vines. And so too has our sexual rapport. We throw ourselves into it with a wild abandon that grows up out of and is nurtured by the fertile soil of familiarity and further encouraged by the intoxicating aromas and exuberance of the season; we play more, we talk while we make love, we experiment, and in doing so pro-long and deepen the pleasure. The once mysterious territory of her body has now been thoroughly staked out, but it contains untold sub-terranean riches that I am only beginning to learn how to excavate.

17 · BEARINGS

WHAT IS HAPPENING this spring in the piazza, in the vineyard, and with Ivana seems somehow all connected, as if these separate spheres of activity are part of a greater whole, revolving in concert with one another, in sync with the rhythms of the natural world as well as the country, region, and culture I have immersed myself in.

This *is* Italy, but it's quite different from what normally comes to mind at the mention of those three magic syllables. This is not your typical Renaissance-painting panorama of terra-cotta farmhouses and cypress trees, of gondolas floating down a canal or *dolce vita* Romans zipping by the Coliseum. Here, far from basking under the Tuscan sun, you are much more likely to get lost in the Piemontese fog, so it took a while for me to get my bearings.

My orientation point, the Piazza del Centro, is in the center of the village of the *comune* (municipality) of Castiglione Falletto, which is located in the middle of the Barolo zone in a hilly area called the Langhe. The Langhe is in the province of Cuneo in the region of Piemonte, way up in the northwest corner of what has come to be known as the country of Italy.

Piemonte is a region on the edge. It borders France, which is a scant two-hour drive to the west from Castiglione, and Switzerland, about two hours to the north, with Germany and Austria not far beyond. Rome, on the other hand, is an eight-hour drive south.

The very name, from the French and Italian words for foot (*pied, piede*) and mountains (*mont, monte*), means "at the foot of the mountains," and that's just where it is: on a clear day, the snow-capped Alps hang surreally on the horizon, shimmering silver-white during the day or glowing red-orange at dawn and dusk. On an *un*clear day—of which there are many—they hover ghostlike in the hazy mist *(foschia)*. Even when they are completely obscured by the thick fog *(nebbia)*, you can sense their presence off in the distance. In the topsy-turvy foothills of the Langhe, you sometimes feel surrounded, as if you're standing in a big bowl of tossed salad with a rim of mountains all around you. But geographical *dis*orientation is only the beginning.

These people don't *look* Italian. Piemontese tend to be tall and fair with the rounded, slightly chunky features of the north—a far cry from my short, dark-skinned grandfather (Ivana's mother would probably have referred to him as *africano*) with his tight-muscled body and sharply chiseled "Roman" nose. They are less open and gregarious than Italians from farther south, and they use hand gestures sparingly.

The area I find myself in is a predominantly rural one (the principal crop happens to be wine grapes), and many of its people are gruff farmer types who generally keep to themselves and don't talk much to strangers. When they do talk they talk differently. Had my grandfather Vincenzo ever bumped into a native Piemontese, the two of them would probably not have had a clue what the other was saying; in fact, most Italians even today would be hard-pressed to understand a conversation between two Langaroli. If this seems a bit strange, you have to remember that the Italian language did not even officially exist until after the Reunification in the 1860s.[1] At that point, the Tuscan dialect—

considered less contaminated by outside influences than the others—was adopted as the official language of the new united Italy (even though many of the founding fathers themselves spoke French). Prior to that time, each independent city-state had its own distinct dialect. Old habits die hard, however, and it was really not until after World War II, when children started going to school in greater numbers, that "standard" Italian began to be more widely spoken and understood.

To this day, Italy remains a land of many dialects—extraordinarily many, considering the size of the country—but most are really just strong accents with a bit of local slang mixed in. Piemontese, however, borders on being a different language altogether. People speak in it, sing songs in it, and recite a litany of proverbs in it. There's even a thick Piemontese-Italian dictionary (Ivana's mother has a copy, which she kindly lets me borrow).

Here in Castiglione, almost everyone speaks Piemontese. It's the predominant—often exclusive—language spoken in Renza's bar, and even those who don't actually speak it understand it because their parents spoke it. I want to speak it too; it seems as typical a product of this place as the wine or food. So, armed with my dictionary, my knowledge of Italian and French, and the impatient tutelage of Renza, I attempt to learn it. But the dialect proves even harder to grasp than a hand of *tarocchi*.

French helps, as many Piemontese words are French in origin, but it will take you only so far. Much of the dialect is indigenous patois and bears no relation to either French or Italian. Take the simple word "onion," for example: it is *oignon* in French, *cipolla* in Italian, and *sciule* in Piemontese. Even worse, it seems as if the dialect-target is always moving: just when I think I've mastered the Piemontese word for rabbit (*coniglio* in Italian) as *pèrru*, with an accent on the first syllable and a quickly flipped double *r*, I hear someone from Dogliani refer to it as *lapin* and someone else from the Roero area call it *cunij*. Four totally different names for the same mute little animal! It makes trying to learn Piemontese frustrating and impossibly complicated, but also demonstrates how people here can pin down with amazing speed and precision whether someone is Piemontese or not and, if so, exactly where within the region they come from.

Aside from the words themselves, there are many grammatical

usages that are a bit peculiar. One day, sitting around during break time with a group of old vineyard workers, I was perplexed by the recurring use of the word *"botte"* in their conversation.

"What's a *botte*?" I asked, innocently.

"It's the time," someone responded. "We realized it is only *mez bot,* and we don't have to go back to work until *doi bot,* so we have a whole hour and a half for lunch today whereas yesterday we didn't stop working until *un bot,* so we had only an hour."

"Oh, I get it!" I said enthusiastically. "It refers to the hours: *Un bot, doi bot,*" I started counting off to unanimous approval, *"tre bot, tre bot e mez, quart bot . . ."*

"NO!" everyone practically screamed. *"Quart URE!"*

Bot, you see, only applies up to three (or three and a half); after that it switches to *ure* (hours) and stays there until after 12:00 *(mes di* or *mesanuit)* when the *bot* return: one, two, three.

"How come?" I asked

"Well, *bot* means to strike, like the metal rod strikes the inside of the church bells. Back in the olden days" (that is, up until about twenty years ago) "people didn't have watches or cell phones, so when they were out in the vineyards, they had to listen to the bells to know when it was time to stop for lunch."

"Okay, I understand that. But they also had to know when it was time to go home to sleep, no? So why don't you say *set bot*?"

"Perchè non è giusto. It doesn't *sound* right the other way."[2]

As if all this linguistic confusion weren't complicated enough, Piemontese pronunciation itself is extremely difficult. The dialect contains many guttural and nasal nuances, and many subtle contortions of the tongue and mouth are required to pronounce the words correctly. (The standard test is *"Doi povron bagné ant l'euli,"* or "Two peppers bathed in oil.") And if you don't pronounce it just right, with just the right delicate inflection on the words, no one seems to have the faintest idea what it is you are trying to say.

When people are not talking, they're often eating or drinking, and in this too the Piemontese are quite different from other Italians. If it's true that you are what you eat, then these people are four-legged beasts walking around on their hind legs. Though only a generation or two ago, many struggled just to get by and meat was something of a luxury, today the generally well-to-do Piemontese are meat-eaters, confirmed carnivores. Vegetarianism is practically unheard of here, and most people consider it some sort of strange disease. It's not that fruit and vegetables don't exist—on the contrary, nearly everyone has a vegetable garden, and fruit trees are plentiful—but vegetables and fruits are mostly relegated to the beginning and end of a meal, like bookends on a library shelf.

Fresh fish is accessible too (the Ligurian coast is a mere two hours away and the huge market at Porta Palazzo in Torino even closer), but this accessibility is a fairly recent phenomenon and the Piemontese are stubborn creatures of habit. With the exception of freshwater trout, eel, anchovies, salt cod, and canned tuna, fish is just not a part of the local cuisine and for most Piemontese remains an occasional and rather exotic treat.[3]

Here in Piemonte, meat rules and beef reigns supreme. But not just any beef. Most of it comes from a certain local breed of cattle called *Razza Piemontese* and, as with Barolo, conformity to the specifications is tightly regulated and protected.[4] Most of these animals are raised by small independent farmers, their breeding is closely controlled, and proof of their provenance is publicly displayed in the shops where it is sold. Butcher shops are numerous here (little Gallo has three of them; even tiny Castiglione has one, though it's far from the center of the village), and the proprietors are all experts at their trade; they get the animals straight from the slaughterhouse and cut them up themselves to the specifications of their meat-savvy clientele.

Because the meat of the *Razza Piemontese* is very low in fat and has a delicate, subtle flavor, it is usually either thoroughly cooked—stewed, boiled, braised in wine, or roasted—or completely raw. Ubiquitous appetizers are Insalata di Carne Cruda (which Silvana prepared at our first dinner) and *Fettine albese* (a k a *alla zingara,* or Gypsy style), thinly sliced raw beef drizzled with lemon, olive oil, salt, and a hint of fresh garlic. People here love raw sausage too (Ivana much prefers raw to cooked), and the nearby town of Bra is justifiably famous for it. It is lean (consisting mostly of ground veal with just a little bit of pork), flavorful, and delicately seasoned with an assortment of spices such as nutmeg, cinnamon, and black pepper—it's almost a shame to "ruin" it by cooking.[5]

As if to counter their hard-edged personalities and thick builds, people here generally prefer their food soft and thin. Pasta is usually thoroughly cooked rather than served *al dente* as it is in other parts of Italy. Most of the bread is soft and airy (hand-rolled *grissini*, on the other hand, are crispy, crunchy, extremely long, and totally addictive), and customers in bakeries will often go even further requesting loaves that are *morbidi* (soft) or *poco cotti* (not too cooked). Unlike thick Tuscan beefsteaks or hefty *frittate* chock-full of chunky vegetables, here *frittate* are usually no thicker than a pancake while the *bistecca* are sliced almost paper thin on a meat slicer. Ask a butcher for a steak *un po' più spessa* (a bit thicker) and you will be lucky to get much past a quarter inch.

But don't worry: whether at someone's home or a restaurant, you are not likely to leave the table hungry.

The Piemontese are famous—or infamous—for their endless barrage of antipasti.[6] Typical antipasti include *vitello tonnato,* Bagna Cauda, trout or eel or zucchini or eggs *in carpione* (cooked and marinated in white wine, vinegar, onions, and sage), *fritatte* (open omelets), boiled calf's tongue or anchovy fillets with green sauce, rolls of sliced ham filled with mayonnaise covered with aspic, warm flan-like vegetable custards, little poached sausages called *cotechini* with mashed potatoes or Fonduta, and *Insalata russa* (a mayonnaisey salad of potatoes, carrots, peas, hard-boiled egg, and canned tuna). Sweet peppers are very popular and turn up frequently, roasted and rolled with a mousse of tuna, perhaps, or drizzled with Bagna Cauda sauce.[7]

After this long parade of antipasti, typical *primi* are mercifully few: Tajarin and Agnolotti del Plin. Both are generally available in one of two ways: *al sugo* (a tomato-based meat sauce) or *burro e salvia* (tossed with melted butter and fresh sage; you can leave out the sage and shave truffles on top in the fall and winter).

While dried pasta made from durum semolina is widely consumed throughout Piemonte, it is a modern import from outside the region. Typically, fresh pasta is made with soft-wheat flour and lots of egg yolks, which give it a rich yellow-orange color. As an alternative to pasta, there are *gnocchi* (potato dumplings), often served in a thick cheese sauce, or risotto (when made with Barolo wine the plump white rice turns a lovely shade of purple).

Besides beef, people eat a lot of rabbit (Renata's mom, Domenica, makes a killer Cuníj all'Arneis, using the indigenous white wine of the Roero, where she is from) as well as chicken—even turkey—and pork. In the fall there is lots of tripe and wild boar, as well as porcini mushrooms and white truffles (for those who can afford them). Polenta is consumed often, especially in the colder months.[8]

There are many local cheeses such as Murrazano, Raschera, Castelmagno and Robiola, which mostly come from the high pastures of the Alta Langa. Gorgonzola turns up often as does Fontina, from the nearby Valle d'Aosta, which is used to make Fonduta. Letting nothing go to waste, scraps of leftover cheese are used to make *brus* (from the Piemontese word for "burn"), a stinky, pungent—some would say lethal—concoction where the cheese is put in a jar (often with a touch of grappa) and allowed to age for months or years. Some purists contend it is not ready until crawling with worms.

Of course, I have known of these dishes for a long time. My friend Bruna, whom I first met in New York in 1989, is a master of them. She bangs out perfect little Agnolotti del Plin, perfunctorily pinching the pasta between her thick strong fingers. When you eat them, the flavor bursts through the silken pasta with a subtle richness and delicacy accentuated by the butter and sage she tosses them in. Her *Tajarin al ragù* sets the standard for all others and her Brasato al Barolo is perfection itself: meltingly tender but not mushy, the meat is imbued with (though not overwhelmed by) the essence of the wine, while the sauce has just the right consistency—not too thick, but not watery either—and a perfect balance of meatiness, sweetness, and acidity.

Regional recipes carry with them a strong sense of the place they come from—of its geography, culture, and history—like a sort of place capsule or edible postcard. But it's totally different to experience regional dishes on their home turf. Here, they are not some unusual, exotic item on a menu or some sort of culinary novelty but an inherent aspect of the collective consciousness, built up over time and passed down through generations. These traditional dishes are part and parcel of the place itself, like the particular quality of the air or light.

*   *   *

Like the language, the landscape, the food, and the people, wine in the land of the Tongues is different too.

Piemonte has its own traditional system of viticulture and nomenclature that is unique to the region. Here a vineyard is measured in *giornate piemontesi* rather than hectares.[9] A *giornata*—3,810 square meters—is thought to have come from the amount of land a farmer and his ox or mule could reasonably cover in a day. Each *giornata* consists of ten subunits called *tavole* (each *tavola* is 381 square meters). Like *vecchie lire,* many people still use *giornate* to quantify the size of a particular vineyard. A large elongated wooden container called a *brenta* was used to measure and transport wine.[10] And a special wine bottle called the *albeisa* was developed to hold it; it is similar to but distinct from the classic Burgundy bottle, brown tinted (as opposed to green), and rather thick and stocky, not unlike many of the winemakers themselves.

More important, Piemonte has the nebbiolo grape. It is indigenous to these hills and is thought to have been named after the dense fog *(nebbia)* that typically cloaks them around harvesttime. The grape's presence here can be traced back at least to Roman times, and it appears to be quite happy right where it is, for, like the Piemontese who are often referred to as *bougia nan* (nonmovers), it has by and large resisted being transplanted anywhere else. Come to think of it, the nebbiolo grape, with its thick skin, light pigment, gruff tannins, persnickety temperament, and understated nobility, seems remarkably similar to the people that grow it. Like many other things around here, it goes by a number of different names and appears in a number of different manifestations throughout the region.[11] But the noblest expression of the grape, the undisputed king of the nebbiolo family (some might argue of wine, period) is the wine called Barolo.

Like other aspects of Piemontese culture, Barolo owes much to the influence of nearby France. While the origins of the nebbiolo grape may go way back into the murky depths of geologic time and the earliest meanderings of the ancients, the birth of this wine is much more recent. In fact, the event can be tightly pinpointed in place and time. Its inception was no accident, and it even has a host of proud parents and guardians, individuals who not only had a hand in creating the wine but who helped shape the history of Italy and France as well.

· RECIPE #9 ·

CUNÍJ ALL'ARNEIS

———————

Rabbit Braised in Arneis, adapted from Domenica Ferrero

INGREDIENTS

1 healthy-sized rabbit, 3 to 4
    pounds (including giblets),
    cut into 9 or 10 pieces
¼ cup extra-virgin olive oil (or
    more as needed)
1 teaspoon butter
Salt and pepper to taste
Enough all-purpose flour to
    dust rabbit (about ⅓ cup)
1 onion, peeled and cut into
    small dice (about 1 cup)
1 carrot, peeled and cut into
    small dice (about ½ cup)
3 ribs of celery cut into small
    dice (about 1 cup)

2 small cloves garlic, smashed
    and coarsely chopped
1 fresh herb bouquet consisting
    of rosemary, sage, basil (if
    available), oregano, and bay
    leaf (1 fresh or three dried),
    tied together with string or
    cheesecloth
Pinch of nutmeg
2 glasses (about 1½ cups) arneis
    (see note below)
1½ cups (1 14-ounce can) chicken
    broth

1. Prepare the rabbit: remove the heart, liver, and kidneys from the inside of the rabbit, and reserve. Cut off the head and discard. ("Some people like to use it for stock," Domenica says, "and others like to cook it and nibble on it, but I don't bother anymore.") Cut the rabbit into pieces; two forelegs, two hind legs, the rib cage split lengthwise in two, and the saddle cut in half or, if the rabbit is especially large, into three pieces. (You may have your butcher do this for you.) Rinse well and pat dry.

2. Heat the oil and butter in a large heavy-bottomed pot. Sprinkle the rabbit with salt; lightly dredge with flour, patting well to remove any excess, and add to the pot in a single layer. Cook the rabbit on a medium-high heat, turning the pieces as necessary so that they brown evenly without burning. (If it is not possible to add all the rabbit in a single layer at one time, do this in stages.)

3. When the rabbit is golden brown, remove to a plate. Add the onions and cook, stirring, until softened and nearly transparent. Add the vegetables and garlic and continue cooking on medium-high heat, stirring occasionally, until they begin to soften. If the bottom of the pot is too dry, add a bit more olive oil. The vegetables may take on just a bit of color but do not let them brown.

4. Once the vegetables begin to soften, add the herb bouquet, a pinch of nutmeg, salt and pepper to taste. Add the browned rabbit pieces (along with the reserved heart, liver, and kidneys) to the pot and mix with the vegetables. Raise the heat to high; add the wine and let cook on high heat until the alcohol is evaporated and the wine has reduced to about ½ cup.

5. Add the chicken broth. When it comes to a boil, lower heat to a simmer, cover, and cook for 1 to 1½ hours, or until the rabbit meat is tender, the stock has thickened, and the vegetables have begun to break down into the sauce. Remove the herb bouquet and discard.

May be made several days ahead and reheated as necessary.

YIELD: SERVES 4 TO 6 AS A MAIN COURSE.

NOTE: Arneis is a white wine made from a grape indigenous to the Roero area across the Tanaro River from Barolo. Brought back from near extinction about a decade ago, the wine is pale yellow, light-bodied, exotically perfumed, and pleasantly acidic. If arneis is not handy, use something else, but be sure to avoid heavy, fruity wines with lots of oak.

· RECIPE #10 ·

FONDUTA

FONDUTA COMES FROM the word *fondere*, to melt, fuse, blend, or merge, and those are all pretty good descriptors of the process of making it: the cheese gradually melts into the milk, and the egg yolks blend into the cheese, helping to enrich and thicken it even more. It is essential that this happen as slowly and steadily as possible for the ingredients to truly meld together; otherwise, the cheese will become impossibly stringy and clumpy and the egg yolks may curdle, in which case there is nothing to do but dump it and start over.

Fonduta is extremely versatile. Typically it is served in the fall and winter with rich, fatty *cotechino* sausages or a delicate vegetable flan or braised cardoons, but it's also great in the spring with artichokes or used to fill ravioli, as I did at my restaurant, which are then tossed with fresh asparagus, butter, and Parmesan cheese.

It makes a wonderful filling for crêpes, but it's also delicious simply spooned over toast rounds.

INGREDIENTS

1½ pounds Fontina cheese
    (preferably from the Valle
    d'Aosta)
1 quart milk

6 egg yolks
1 teaspoon butter
½ tablespoon flour

1. Remove the orange-colored rind from the cheese. Cut the cheese into little pea-sized pieces and put them in a large stainless steel mixing bowl. Pour the milk over the cheese (there should be just enough to cover), cover the bowl with plastic wrap, and let stand at room temperature for two hours or overnight in refrigerator. (Note: Letting the cheese and milk sit together for a period of time will help them blend more easily.) When you are ready to prepare the Fonduta, remove from the fridge and let warm to room temperature.

2. Select a pot with a diameter large enough to form a double boiler with the mixing bowl. Fill the pot halfway with water, place over high heat, and bring to a boil. While the water is heating, add the egg yolks to the cheese and mix in with a whisk. Add the butter in one piece; sprinkle the flour over the top of the cheese and mix in.

3. Lower heat under the pot so that the water is on a steady simmer. Place the mixing bowl over the pot and begin whisking. *(See note below.)* Whisk frequently while the cheese slowly melts, removing the bowl from time to time if it gets too hot (it should never boil). Whisk down to the bottom of the bowl to make sure that it is not curdling,

and scrape the bottom and sides from time to time with a rubber spatula.

4. First, the cheese will melt and liquefy; then it will thicken. In the end (after about 30 to 40 minutes) it will resemble a thick creamy sauce; it should be pale yellow in color, totally homogeneous and not grainy. The finished Fonduta should have the consistency of a thick hollandaise.

5. Use immediately as a component of a hot appetizer. Allow any leftovers to cool. Then refrigerate and use as a filling.

YIELD: MAKES ABOUT 2 QUARTS OF FONDUTA.

NOTE: Some experienced cooks make Fonduta directly in a heavy-gauge stainless steel pot. While this method is equally effective (and much quicker) than the one above, the chance of the mixture getting too hot and curdling is also much greater. For this reason I suggest using the double boiler method.

· RECIPE #11 ·

PASTA FRESCA ALLA PIEMONTESE

Fresh Pasta, Piemontese Style

IN PIEMONTE, PASTA is usually made with a type of soft-wheat flour called "Tipo OO" and an abundance of egg yolks. (Some cooks use all yolks, but I generally prefer to use 2 whole eggs to every yolk.) This gives the pasta a distinctive pale orange color (Italian eggs are very orange, to begin with) as well as a richness and delicacy in the mouth. You can simulate "OO" flour by mixing 1 part all-purpose with 3 parts pastry flour.

Making fresh pasta is simple and quick. It is actually more of a *process* than a recipe. The most important thing is to know the final result you're going for; then listen to the dough in your hands and it will tell you what to do. After making it a few times, you too will be a fresh-pasta pro. Preparing the dough a day ahead will help streamline the actual pasta preparations.

INGREDIENTS

1 pound flour (or more as needed), preferably "Tipo OO" or a blend of all-purpose and pastry flours	*4 whole eggs* *2 egg yolks* *1 scant teaspoon salt* *1 tablespoon extra-virgin olive oil (optional)*

1. Place the flour on a work counter and make a well in the center. Crack the eggs and egg yolks into the well. Add the salt and olive oil (if you are using it) and mix the eggs with a fork.

2. When the eggs are well blended, begin to mix in flour from the inside edge of the well, little by little, making sure it is thoroughly absorbed by the moisture of the eggs before adding more. This will help prevent lumps from forming.

3. Continue gradually incorporating flour, always from the inside edge of the well and always concentrating the mixing in the center. When it becomes too difficult to use the fork, continue with the hands. (Note: you are still gently mixing flour in, not kneading.)

4. When the dough begins to take shape but is still quite soft and moist, transfer it to another area of the worktable. Using a pastry scraper, scrape the remaining flour up off the surface and pass it through a fine sifter onto the dough. Discard any bits of dough remaining in the sifter.

5. Continue working flour into the dough until it forms a soft ball. Move any remaining flour off to the side and begin to knead, keeping the surface well dusted with flour. Continue kneading until the dough is firm and resilient (about 5 minutes). Poke the ball of dough: it should bounce right back, and the interior should be only mildly sticky.

6. Lightly dust the dough with flour, wrap with plastic, and let rest for at least 1 hour or overnight in the refrigerator.

YIELD: MAKES ABOUT 1½ POUNDS OF PASTA DOUGH.

· RECIPE #12 ·

AGNOLOTTI DEL PLIN

Pinched Ravioli, adapted from Bruna Alessandria

PLIN MEANS "PINCH," and, as you will see below, that is exactly how they're made. If properly done, these little pasta dumplings resemble tiny oblong pillows or odd hats with their brims folded up on one side. It is a unique and wonderful shape, especially combined with delicate, paper-thin pasta and the subtly sweet, meaty filling encased in it, and so ubiquitous that the terms *agnolotti* and the more generic *ravioli* are used interchangeably throughout the Langhe area.

Some people prefer their *agnolotti* sauced with tomato-based meat *ragù* or a rich concentrated gravy called *sugo di arrosto,* while others serve them *al fumo* ("smoking"), with nothing but the steam rising from them, to prove how delicious they are. Most often, they are lightly tossed with butter and sage and sprinkled with a little grated cheese. If you happen to have some fresh truffles around, that would be fine too.

Because making *agnolotti* is a rather long and involved production, it makes sense to make a big batch of stuffing, divide it into small packets and freeze some for later. (If you prefer to make a smaller quantity, simply divide the recipe in half, or, if you wish, omit the rabbit and adjust the quantities of other ingredients as necessary.) The whole process can be greatly simplified by breaking up the various steps: roast the meat and cook the rice and cabbage one day, pick the meat and make the stuffing on another; make the pasta ahead of time and then invite friends over for a fun-filled afternoon of pasta-pinching followed by dinner. Freeze any excess ravioli and/or stuffing for future use.

INGREDIENTS

FOR THE FILLING

½ cup arborio rice

1 teaspoon coarse salt

½ cup plus 2 tablespoons olive oil

1 tablespoon butter

1 cabbage (preferably Savoy), quartered, center core removed, and sliced

1½ pounds of pork from the leg, loin, or shoulder cut into large pieces

1½ pounds of veal, cut into large pieces

1 small rabbit, about 2½ to 3 pounds, cut into pieces

2 small carrots, washed, split, and halved

3 ribs of celery, washed and halved

1 large onion (or 2 medium), peeled and quartered

1 bunch fresh rosemary

3 cloves garlic, lightly crushed

FOR THE RAVIOLI

1 pound of meat filling

2 eggs

Pinch of nutmeg

Grated Parmigiano Reggiano or Grana Padana cheese

Salt to taste

1 pound of fresh pasta dough (see recipe page 104)

Semolina flour

TO SERVE

¼ pound butter

1 bunch fresh sage, leaves picked

Salt to taste

Grated Parmigiano cheese

I. PREPARE THE FILLING

1. Cook the rice: put the rice in a pot, cover with 3 cups water, bring to a boil, and simmer until cooked (about 20 minutes). Add a pinch of salt halfway through. Strain the rice, rinse with cold water, and set aside.

2. Cook the cabbage: heat 2 tablespoons olive oil and butter in a large frying pan or shallow saucepan. Add the cabbage, a pinch of salt,

and stir. Add two tablespoons of water to the pan, cover, and cook until the water has evaporated and the cabbage is soft (approximately 20 to 30 minutes). Let cool, squeeze out any excess moisture, and set aside. (Note: Both the rice and cabbage may be cooked a day or so before.)

3. Cook the meat: preheat oven to 450 degrees. Place the meat, carrots, celery, onions, rosemary, and garlic in a large shallow roasting pan. Drizzle with ½ cup olive oil, sprinkle with salt, and toss to coat. Roast in the oven, turning occasionally, until the meat is browned (about 30 minutes). Lower the temperature to 375 degrees and continue roasting, stirring occasionally, until the meat is tender, about 1 hour.

4. When the meat is cooked, remove the roasting pan from the oven, transfer the contents to a large bowl, and let cool. Clean the rabbit meat off the bones. Remove the vegetables and rosemary, leaving only the cooked meat in the bowl.

5. Add two fistfuls of cooked rice to the bowl and about two-thirds of the cabbage. Mix together, then pass the contents through a meat grinder (if you don't have a meat grinder, you may also pulse in the bowl of a food processor or chop by hand). The ground meat should have the consistency of a dense, stiff paste. If it seems too dry and grainy, add a bit more of the cabbage and pass once again through the grinder. (The cabbage adds moisture to the filling while the rice adds substance. Quantities of both may need to be adjusted depending on your preference and the meats you're using.)

The filling may be made ahead, divided into small packets, and frozen until ready to use.

YIELD: MAKES ABOUT 4 POUNDS OF FILLING.

II. PREPARE THE RAVIOLI

1. Place about 1 pound of filling in a bowl. Add 1 whole egg and 1 egg yolk (reserve the leftover egg white), a pinch of nutmeg, 1 tablespoon of grated cheese, and salt to taste, and mix well. The filling should now be slightly moister than before, and, though still fairly stiff, homogeneous (you may still see some little white bits of rice) and smooth. Transfer the filling to a pastry bag with an open tip.

2. Using a rolling pin or pasta rolling machine, roll the pasta out

into a long sheet as thin as possible and about 6 inches wide (or, if using a pasta roller, as wide as the width of the machine); it should be almost transparent. Dust the pasta with flour to prevent it from sticking, but make sure the top-side of the sheet is free of any excess flour.

3. Lay the sheet out on a clean, lightly floured work surface. Working quickly to prevent the pasta from drying out, pipe little dabs (approximately ½ teaspoon) of filling in a straight line along the entire length of the sheet about ½ inch above the lower edge. The dabs should be close together, with just enough space between them to fit your fingertips.

4. Dip a pastry brush in the reserved, lightly beaten egg white and brush in a long line just above the dabs of filling.

5. Carefully lift the entire lower edge of the pasta sheet, fold it snugly over the filling, and tamp the edge down firmly on the moistened sheet of pasta.

6. Using thumb and forefinger, pinch the pasta together in the space between each of the little mounds of filling, working your way down the length of the sheet, to form individually enclosed mounds of filling.

7. Using a pastry roller-wheel with a scalloped edge, cut down the length of the sheet just above the filling (making sure not to cut into the filling itself) and pull the strip away from the rest of the pasta. Pressing down hard with the pastry wheel, make short quick cuts on each pinch to form the individual ravioli.

8. Transfer the *agnolotti* onto a wooden pastry board or sheet pan covered with parchment paper, and dust with semolina flour, making sure they are not too crowded together. Repeat the process with the rest of the pasta sheet.

The *agnolotti* may be made ahead and frozen on baking sheets, then in small plastic bags until ready to use.

YIELD: MAKES ENOUGH RAVIOLI FOR 6 TO 8 PEOPLE AS AN APPETIZER OR 4 PEOPLE AS A MAIN COURSE.

III. TO SERVE

1. Bring a pot of salted water to boil. Add the pasta and cook until done (the *agnolotti* will usually cook in about 2 to 3 minutes; when they rise to the surface, they are done).

2. Meanwhile, melt the butter in a saucepan, and add the sage leaves and a pinch of salt.

3. When the butter begins to sizzle, add a spoon or two of water from the pasta pot.

4. Strain the cooked *agnolotti*, add to the butter and toss.

5. Divide the pasta into individual bowls or a large platter (leaving the sage leaves in with the pasta), and serve, along with grated cheese.

NOTE: This filling likely evolved as a way of using leftover roasted meats. You can use pretty much any cut or combination of meat you like; if, for example, you prefer not to use rabbit, substitute chicken or just use more pork and/or veal. Pork, however, is an important component as it is fattier and more flavorful than the other meats.

While some people finely chop the sage, others prefer to keep the leaves whole so that the herb perfumes the pasta without overwhelming the flavor.

· RECIPE #13 ·

BRASATO AL BAROLO

Beef Braised in Barolo, adapted from Bruna Alessandria

INGREDIENTS

¼ cup olive oil

1 piece of beef such as chicken steak (see note below) about 3 to 4 pounds, tied

2 medium onions, peeled and cut into pieces

2 carrots, peeled and cut into pieces

5 ribs celery, washed and cut into pieces

3 or 4 cloves garlic, smashed

Salt and fresh ground black pepper to taste

3 sprigs of rosemary

2 fresh bay leaves (or 1 dried leaf)

6 glasses (about 4½ cups or 1½ bottles) Barolo (see note below)

2 cloves

6 whole black peppercorns

1 cinnamon stick

Water, veal stock, or chicken broth

1 small piece sweet butter, chilled (optional)

1. Heat the oil in a wide, high-sided, heavy-bottomed pot or rectangular roasting pan just big enough to comfortably hold the meat. Add the chicken steak and cook on a high heat until well browned on all sides.

2. Add the vegetables and garlic, and season with salt and pepper. Tie the rosemary and bay leaves together with string (this way they may be easily removed later), add to the pan, and continue cooking until the vegetables too begin to turn golden brown (about 10 to 15 minutes).

3. Add the wine, the cloves, peppercorns, and the cinnamon stick. (There should be enough wine to almost cover the meat.) Bring to a boil and cook on high heat for about 5 minutes. Lower heat to a gentle boil and simmer until the wine is reduced by about half (35 to 45 minutes), occasionally turning the meat. By this point, most of the alcohol should have cooked off and the wine will have lost its raw acidic edge.

4. Add enough water or stock to just barely cover the meat and bring to a boil. Lower heat to a simmer, cover, and continue gently braising for 1½ to 2 hours, turning occasionally, until the meat is thoroughly cooked (a knife inserted into the center should go in easily).

5. Carefully remove the meat from the pot, allow it to cool, and then refrigerate. The remaining contents of the pot should be about one-third liquid and two-thirds vegetables; if it is too liquid, cook on a high heat until the excess liquid has evaporated. When the vegetables are cool, remove the herb bouquet and discard. Pass the entire contents of the pot through a food mill fitted with the small-hole die. Taste for seasoning, adding additional salt and pepper if necessary, and refrigerate. (May be made several days ahead up to this point.)

TO SERVE

Remove the string from the meat and cut into slices about ½ inch thick. Arrange the pieces in a single layer in a large frying pan, add enough sauce to cover the meat, cover, and cook gently until the meat is thoroughly heated through. (If the sauce is too thick add a little broth or water; it should be thick but spoonable and not ooze excess liquid.) Transfer the meat slices onto plates with a slotted spatula. Add

the cold butter to the sauce (if you are using it), and stir to dissolve.
Spoon the sauce over the meat and serve.

YIELD: SERVES 6 TO 8 AS A MAIN COURSE.

NOTE: The preferred cut for this preparation is called *tenerone* (big tender) in Italian. In America, I like to use a cut called chicken steak, a
cylindrical piece from the neck. It is tender, stringy, and full of flavor—
perfect for long, slow braising; even Bruna was impressed. It has a ge-
latinous tendon running down the middle that should be removed
before tying and cooking. If chicken steak is not available, ask your
butcher to recommend another cut suitable for braising.

A NOTE ABOUT THE WINE: This dish, when done properly, is a perfect marriage of meat and wine. During the long cooking process with
the meat practically submerged in the wine, the two principal ingredients become one—the wine disappears into the meat and the meat
turns nearly black with the wine. For this reason, the wine you use is
very important to the end result. Some cooks here use Barolo, both because they have it and to make a good impression (*fare una bella figura*).
Many, however, while still calling it Brasato al Barolo, use nebbiolo
wine instead, and you should feel free to do so too. (I myself would
rather save the Barolo to drink with dinner.) If nebbiolo is not available, by all means use something else. Just make sure it is not too
jammy, fruity, and concentrated, and doesn't taste too much of new
wood. The ideal wine would be medium to full-bodied, with pronounced grapey flavors, moderate acidity, and firm tannins.

· RECIPE #14 ·

BAGNA CAUDA

Hot Bath

DURING FALL AND winter, the distinctive aroma of bubbling garlic
and oil pervades the Langhe. Many Langaroli say it increases resistance

to viruses and acts as an effective prevention against cold-weather dis-
eases. Many also claim it bolsters sexual performance, cautioning,
however, that it only works if both parties partake.

For this dish garlic, which does not usually play a big part in the cooking of Piemonte, is thinly sliced, blanched in boiling water (some cooks
use milk), and then slow-cooked in simmering oil until it literally disinte-
grates, blending with the oil to resemble a loose, chunky puree. The long
cooking process takes away the powerful, acrid punch of raw or quickly
sautéed garlic. In more rustic, traditional preparations, parboiling is dispensed with and a hefty fistful of anchovies is added, turning the *bagna*
almost brown and giving it a pronounced blast of pungent salty fish.

Bagna Cauda is typically served in special raised terra-cotta bowls
with space underneath for a candle to keep the sauce warm while it is
being consumed. The sauce is served with a selection of raw and partially cooked room-temperature vegetables, which are dipped into the
hot oily sauce. When the vegetables are finished, whatever's left is
soaked up with a piece of bread. Some people like to crack an egg into
the bowl and scramble it with the last bits of garlic and oil.

INGREDIENTS

*4 cups fresh garlic, peeled and
    thinly sliced (see note
    below)*
*3 cups extra-virgin olive oil
    (plus more, if necessary)*
*8 anchovy fillets (use more or
    less, according to taste), cut
    into large pieces*

*1 tablespoon butter
    (optional)*
*1 egg per dish
    (optional)*
*Assorted raw/cooked vegetables
    (see note below)*

1. Bring water to boil in a heavy-duty saucepan; add the sliced garlic to the boiling water, wait about five seconds, and then pour out
through a strainer.

2. Place the strained (but still warm) garlic back into the empty (but
still warm) pot, pour the olive oil over the garlic—there should be
enough oil to just barely cover the garlic, if not, add a bit more—and

return to the stove over a very low flame. It is important that the garlic heat up slowly together with the oil.

3. Cook the garlic slowly over low heat. Some bubbles should come rising up to the surface, but the oil should never come to a full boil. Stir occasionally with a wooden spoon until the garlic begins to soften and break down into the oil (about 45 minutes to 1 hour). Make sure the garlic does not burn.

4. Add the anchovy pieces and continue cooking, stirring occasionally and mashing any remaining chunks of garlic, until the anchovy has broken down into the sauce (approximately 20 minutes). By this point, the garlic should be homogeneous, soft, and fairly well blended with the oil. There should be only a thin layer of excess oil on top.

5. Add the butter (optional) and stir in to dissolve. This should help amalgamate it even more.

6. Carefully spoon the Bagna Cauda into terra-cotta dishes, light the candles underneath, and serve along with an assortment of vegetables.

YIELD: SERVES 6 TO 8 AS AN APPETIZER.

Bagna Cauda may be made several days in advance, stored in a sealed container in the refrigerator, and heated up as needed. Just make sure that the garlic is covered by a thin layer of olive oil.

————————

NOTES: The garlic should be sliced against the grain—that is, across the diameter, not lengthwise; this allows it to break down faster and blend more completely into the hot oil. For the same reason, if the garlic you are using is mature and has a large shoot in the middle, this should be removed as it is much tougher and will take longer to dissolve. Also, be sure to discard the dark tips at the root end of each clove.

Following is a list of vegetables typically served with Bagna Cauda but, by all means, use whatever you like and is available. All are raw unless otherwise noted. The vegetables should be peeled and seeded, if necessary, and cut into appropriately sized pieces to pick up and dip:

Red and yellow sweet peppers, seeds removed
Savoy cabbage (use the tender inner leaves)

Jerusalem artichokes (called *topinambour* in Piemonte, they are some-
   times called sun chokes in America)
Celery (or cardoons if you can find tasty tender ones)
Fennel
Radishes
Scallion tips (may be a bit overdoing it!)
Beets, cooked, cooled, peeled, and cut into pieces
Potatoes (prepared like the beets)
Belgian endive leaves
Broccoli rabe (lightly blanched; not traditional but I like the bitterness)
Cauliflower

18 · A TALE OF TWO CITIES

BAROLO IS KNOWN as the Wine of Kings and the King of Wines, an
epithet with all the subtlety of a trumpet fanfare. But it is not merely
braggadocio. Barolo has many royal connections; Piemontese nobles
(including the first king of Italy) were among the first to produce it
and they all drank it. Plus, there is the aristocratic quality of the wine
itself.

However you account for the wine's regal character, one thing is
clear: every king needs a queen, and Barolo has one. Her name is Giulia. Actually, she (like Ivana) has many names. Juliette Françoise Vic-
tornienne Colbert de Maulévrier was born in France, in the Vendée
region of Poitou, in 1786. Despite the good fortune of being born into
a wealthy aristocratic family,[1] her life, especially her childhood, was not
a particularly pleasant or carefree one.

Her mother died when she was only three years old, on July 14,
1789—the same day the Bastille was stormed—and most of her relatives perished under the guillotine in the bloody revolution that fol-

lowed. Juliette survived only by sneaking out of the country disguised as a boy along with her father and two brothers.

Little Juliette was a prim, proper, pious, delicate, and rather serious child by nature. But with the loss of her mother and forced exile, these qualities intensified. At a very early age her father became her private teacher as well as her sole guardian. He provided his daughter with a strict, well-rounded classical education—quite unusual at that time for a female, even among the privileged classes. She studied French, German, Italian, Latin, philosophy, geography, history, physics, and mathematics and showed a surprising aptitude for them all.

After narrowly escaping the bloodbath in France, she and her remaining family shifted around northern Europe, safe but nevertheless displaced foreigners, without the comforts and connections they had enjoyed prior to the Revolution. In 1804, Napoleon came to power, pardoned noble exiles, and guaranteed their safety if they returned to France. The Maulévriers immediately accepted the offer and went home. Upon her return, Juliette, now a young woman, assumed her rightful role as a demoiselle of the upper class, becoming a fixture in Napoleon's court in Paris. It was nowhere near as opulent and excessive as the court of Louis XVI had been, but it was still a dazzling place to be. Wit and intelligence had taken the place of the previous royal shenanigans. The king and his powdered consort had fallen; a dashing general (noble but not royal by birth) had ascended to the imperial throne, ushering in a whole new world order in which most anything seemed possible. Though a monarchist like her father, Juliette was smart, attractive, and proud, and felt right at home in this milieu.

At court, Juliette encountered many extraordinary people—nobles, diplomats, scholars, and soldiers from all over Europe—but one stood out from the rest; a young Italian count named Carlo Tancredi Falletti di Barolo. Actually, their meeting was not mere chance; it was conceived by Napoleon himself (the emperor frequently forged amorous alliances between individuals to solidify political ones), and shortly thereafter a marriage was arranged by Camillo Borghese, an Italian nobleman who was governing Piemonte for the French.[2]

Torino at that time constituted a sort of northern capital of the Italian peninsula, a little sister city of Paris. Here, French was the language of the upper classes, and cultural cues were taken from nearby France and Austria. As a young ambitious noble, Carlo had been sent to Napoleon's court, where he began as a page but quickly worked his way up to court chamberlain.

Most arranged marriages are just that—an *arrangement* motivated principally by convenience, necessity, or the prospect of political, financial, or social gain. From this perspective, Napoleon had made a great match. The Falletti were one of the wealthiest families in Europe and well established in Piemonte, which was strategically located at the center of the European axis of power; Paris to the northwest, Vienna to the northeast, and Rome to the south.

But this union turned out to be much more than anyone could have anticipated. Carlo and Giulietta were perfect for each other; both were highly intelligent, conscientious, and sensitive, as well as extremely religious, yet also firmly grounded in the real world. Even their differences were complementary: Carlo was more cautious and conservative; he studied an issue from all angles and considered all the possible ramifications before moving forward, while Giulia, perhaps because of her tumultuous childhood, was more impulsive and emotional.

In no time the two forged an unshakable bond that was to last their entire lives. On August 18, 1806, in a lavish though not excessive ceremony, the marriage of Carlo and Giulietta—man and woman, Italy and France, reason and emotion, piety and politics—took place in Paris. When it was over, Giulietta added yet two more names to the many she already had: Falletti di Barolo.

In the early years of their marriage, the couple divided their time between Torino and Paris. These two cities were the twin capitals of the new Europe and both were undergoing major transition: post-Revolution Paris was still in the process of reinventing itself while over in Torino rumblings were beginning that would soon erupt into the struggle to reunify Italy. In both cities the young Marchesi di Barolo established important salons that were frequented by the most influential men and women of arts, letters, science, and politics in Europe. Then in 1814, things changed once again: Napoleon's reign ended, the fallen emperor went into exile on the island of Elba, and the couple settled once and for all in Piemonte. With that, Juliette, whose name had been shifting back and forth between French and Italian for much of her adult life, became decidedly Giulia.

By this time it was apparent that the loving couple, though utterly devoted to one another and blessed in so many ways, were not to have children. In the absence of offspring, the two turned first inward, then outward. Their piety deepened as did their social activism. With no direct heirs, they bestowed their substantial love and social beneficence, financial resources, and social influence on the world around them. And what was around them was, like many urban centers then as now, a meeting place of the very best and the very worst. Unlimited wealth met dire poverty, gilt met grit, and beggars died under the porticoes of the elegant boulevards.

Carlo focused mainly on politics (he was mayor of Torino from 1826 to 1829), while his wife turned her attention to the indigent poor, especially women. Palazzo Barolo in the Via Orfane soon opened its huge wooden doors to the city's destitute and distributed hundreds of meals each day to the hungry.[3] But Giulia didn't stop there. Shortly after settling in Torino, she (along with a group of other wealthy, influential residents) went on a tour of the city's civic institutions that was organized by government officials. The tour included a visit of its prisons, and this experience made a deep and lasting impression on her. Giulia was appalled at the miserable conditions, at the utter hopelessness of the inmates, and at the ineffectiveness of the penal system, and decided to do something about it. In 1821, Giulia submitted a proposal to the Secretary of State for Prisoners for a total reform of the prison system. He accepted her proposal, on the unusual condition that she herself implement it. Giulia accepted the challenge and went on to spearhead the creation of the first-ever all-women's prison, called Sforzate, where "penitent women who had gone astray" were treated with kindness and respect, given a clean, safe place to live, taught a trade, and received help finding a job and a suitable husband. This was a revolutionary new concept.

In 1834, a cholera epidemic hit Torino; almost everyone who could (including Carlo) left town, but Giulia insisted on staying there with "her women." Carlo died four years later, leaving everything—the vast Falletti holdings and administration of their numerous charitable activities—to his wife. Thereafter Giulia began spending more and more time in the family castle in the country. In many ways it reminded her of her early childhood in the Vendée, eliciting fond memories, which she began to put down on paper.[4] All the same, by this time Giulia had left her past life in France firmly behind: she learned the local Piemontese dialect so she could communicate directly with her tenant farmers and they in turn (while continuing to address her as *La Signora* or *La Marchesa*) affectionately whittled her long string of names down to two: Giulia di Barolo.

In the pastoral hills of the Langhe she discovered that poverty and desperation were not the sole provenance of big cities. The Falletti were the most extensive landholders in the area. They had woods, fields of grain, orchards, livestock, and vineyards. Typically, peasants lived on the noble's land and worked it in return for a share of the yield. Most were subsistence farmers who barely eked out a living, received no education whatsoever, and suffered endless illnesses due to the deplorable sanitary conditions. Giulia immediately set about trying to improve the lot of the peasants while at the same time making the extensive Falletti agricultural enterprise more efficient and productive.

One of their principal activities was grape growing and wine making, but the wine was nothing to get excited about: it was weak and insipid, pink, slightly fizzy, and obnoxiously sweet because the spontaneous fermentation of the late-ripening nebbiolo grape was never fully completed. Now, Giulia was no oenophile, but she *was* a native of France, country of the wine-making monks, where wine had long been made through a more controlled and sophisticated fermentation process. Giulia had become familiar with French wine at Napoleon's court, and it was still the preferred beverage of sophisticated Italian nobles back home.

Once she settled in Barolo, Giulia looked critically at the wine her estate was producing and was appalled: she thought they could do much better; and before long, through her insistence and encouragement, they did.

People in the vicinity, landowners/producers and consumers alike, responded enthusiastically to the improved wine. At a dinner, the king of Piemonte, Carlo Alberto, who had a castle and hunting preserve in nearby Pollenzo, tasted it and liked it. Afterward Giulia, consummate hostess that she was, sent a caravan of 325 barrels to the royal palace in Torino, one for each day of the year minus the forty days of Lent. The king was hooked; in 1838, he purchased a castle and vineyards in the village of Verduno, not far from Giulia's castle in Barolo, and enlisted the help of an Italian enologist, General Francesco Staglieno, to oversee the wine-making activities.[5]

Giulia's neighbor across the valley in the Commune of Grinzane, Count Camillo Benso di Cavour, was also in the process of upgrading his estate's wine making.[6] In 1843, he retained the services of a noted French enologist, Louis Oudart. Oudart's unconventional work in Grinzane soon showed impressive results, and Giulia lured him across the valley to overhaul and supervise her estate's wine-making activities as well.

Here, among the Falletti's more extensive vineyard holdings and superior sites along the sloping tongues of earth at the base of the Barolo castle, Oudart focused on coddling the quirky nebbiolo vines (identification of superior sites, introduction of new pruning methods, and orientation of rows to facilitate harvest and maximize exposure) and carefully controlling the vinification process. The outcome was striking: through a combination of French technology and know-how, Italian soil, a unique indigenous grape variety, the vision of a few passionate noble landowners, and the sweat of countless peasants, a new wine was born. It was comparable in quality to foreign imports but unabashedly made on Italian territory, the same territory that was painfully giving birth to a whole new political entity. And it proudly took its name from the place it came from: Barolo.

Giulia was many things in her life—exile, demoiselle, political activist, prison administrator, society matron, feminist, philanthropist, author, marchesa. She liked to get dressed up and hobnob with her noble peers; she enjoyed splendid palaces, delicious dinners, and an occasional crystal goblet of wine. But she took none of these to excess. For her, entitlement was balanced by a profound awareness of human suffering and a compulsion to take whatever direct action she could to make things better for those less fortunate. She formed two religious orders, the Sisters of Saint Anne and the Children of Christ the Good Shepherd, and a charitable organization called Opera Pia, all of which are extant to this day. Upon her death in 1864, Giulia was dressed (according to her wishes) in the habit of a Franciscan nun and interred in the Church of Santa Giulia in Torino, which she had had built, and she is currently on the track of being officially declared a saint.[7]

From the distance of history, Giulia appears a reserved, ascetic person—passionate within strict boundaries, decidedly unhedonistic, perhaps even a bit puritanical. But appearances can sometimes be deceiving.[8] In any case, though Giulia may seem an unlikely candidate for queen (especially of a wine), for lovers of Barolo she will always be first and foremost the queen of the King of Wines.

19 · KING OF THE KING OF WINES

HIS "CASTLE" IS only a stone's throw away from Giulia's, in a modest yellow building with a faded PACE flag hanging from the balcony. On the stuccoed wall next to the anonymous wooden door is a small plaque—CANTINA MASCARELLO—under which is a tarnished little brass button. Push it and wait: eventually Franca, stone-faced "lady-in-waiting," will come padding down the hall, tentatively open the door, and greet you with a quizzical, rather dubious look. "Is he available?" you ask. Wordlessly, she will point you into a little room around the corner and disappear. (She has seen this all before, many times; people come and go, looking for wine, paying homage, and for what? Franca has much to do.)

The room is cluttered with bottles and books and photos. Gray light slants in from the pair of small windows looking onto the street. Directly before you is a big desk that both dwarfs and supports the frail figure seated behind it. His upper half seems to hang on the flat surface while his lower half, obscured, rests in a metal throne on wheels from which useless legs dangle. Thick dusty spectacles seem too large for the shrunken face, delicate long-fingered hands are spotted with age, and the narrow head of thinning white hair is crowned with a faded blue beret. His very presence, however, overcomes appearances, turns them around, as a karate master turns around the negative force of his attacker.

This is Bartolo, King of the King of Wines. Should you need proof of his noble lineage, just remove the blunted sword from the middle of his name like a knight pulling Excalibur out of the rock, and there it is—Barolo.

Bartolo is an unlikely monarch; in fact, he doesn't even admit to being one: "I am a *contadino,* a *vignaiolo,*" he says. But to his faithful subjects, a cult of eno-fanatics who have been fortunate enough to drink his wine, this humility only confirms his eminence.

Actually, he is something of a quiet fanatic himself. He offers a guest a glass of Barolo from an open bottle on the desk while he continues drinking tepid water out of a battered teacup and explains: "I just continue to do what I always have done, make wine as my father did, the way he taught me to do it. I am a traditionalist, not a conservationist: NO BARRIQUE!" he rails. "And no Berlusconi!!"[1]

Bartolo is a bit calmer, if none the less passionate, when the conversation turns back to Barolo. He might be wheelchair-bound and confined for most of his waking hours to this drab little office, but when he talks about wine, his eyes sparkle and his broken body seems to lift up out of the chair, out the window, and fly Chagall-like through the grape-scented air of the Langhe, back to patches of the countryside he knows like the back of his papery hand.

He hasn't actually been to these places for years, but that doesn't matter; the images remain vividly fixed in his mind. Many of them he has put down on paper, turning them into colorful labels for his wine. As he flips through a stack of them (as much for his own pleasure, it seems, as that of an admiring guest) he relates stories about the pictures—people, events, bits of the landscape that inspired them—like an aged Casanova fondly recalling intimate details of his past loves.

He speaks lovingly too of what takes place in the cantina: "Ah, the mystery of fermentation! The sugar turns to alcohol and makes wine. We don't know how or when. We are a small cantina; we don't have those fancy temperature-controlled tanks. Look around; this is not a laboratory. There is no sophisticated technology here. We are wine-makers, not scientists. We just crush the grapes and let it happen."

"The must," he continues, "is like a woman." (As if on cue, Franca pokes her head in the door.) "It has a mind of its own. You can try and make it do something, make it behave when and as you like, but it does what it wants anyway." (Franca raises her eyebrows and tilts her head, emits a small gust of air from her tightly pursed lips before disappearing again; Franca has heard this all before and doesn't have time to stand around listening to nonsense.) "Better just to leave it alone, let it take its time without pushing. Sometimes it takes ten days to ferment, sometimes three weeks; sometimes in weak vintages the grapes need a little help."

He tells how, in the olden days, savvy winemakers would sometimes add grains of sulfur to the vats of crushed grapes to eliminate some of the "bad" yeasts and encourage the "good" ones to go to work on the sugar. Sometimes, if it was too cold, they had to build fires in the cantina to try and raise the temperature enough to induce fermentation (now, if necessary, at Cantina Mascarello they resort to a little portable heater).

These days Bartolo, long-suffering from a degenerative spinal cord condition, has turned much of the work over to his daughter, Maria-Teresa, who enthusiastically carries on the Mascarello mission ("I just do what my father did and my grandfather; I am a winemaker," she says, "a traditionalist, not a scientist!") But each season, from the confines of his office, he relives the roller-coaster ride of harvest and the uncertainty of fermentation as a spontaneous, unexplainable, and uncontrollable thing. "Beh, you never know," he says. "Each time it's different. And it's always a mystery."

Moistening his lips, he looks up from his reveries and out the window at the steady stream of tractors and tourists passing by. "I remember a time—and it wasn't that long ago, you know—when there were mules out there pulling wagons instead of tractors; tourists were nonexistent. Barolo was just another sleepy little village. It was a lot quieter back then, you can be sure."

Bartolo pauses to take a sip of warm water. "And you, who are you?" I remind him that we have met before and mention that I have lately been helping Fabrizio in Le Munie. "Ah yes, Mascarello. Nice piece, that one. I knew his father, Francesco. Yes, we have the same last name but we're not actually related, not that I know of anyway. But of course, I knew him well. He was an excellent winemaker, a real *contadino*. And quite a dancer too! Fabrizio's working in the country now? That's good. Good for him to get out of that stuffy office in Alba and into the vineyard." (Once again he seems to rise up and make a break for the window.)

Despite his near-legendary reputation, Bartolo has stubbornly resisted the temptation to capitalize on it. He has refused to expand his business and increase production, as so many others have done, choosing instead to remain a small, artisanal producer, a living relic of a time gone by when that was the norm rather than the exception. Cantina Mascarello is miniscule from a commercial standpoint. They don't buy grapes from anyone and have the same five hectares of prime vineyard Bartolo's grandfather had, from which they produce approximately thirty thousand bottles of wine each year, about half of it Barolo. They share a portable bottling line with another producer nearby, Giuseppe Rinaldi, who also happens to be a cousin.

One of their larger parcels (a little over a hectare) is in the renowned Cannubi vineyard, but they have steadfastly declined to play the *cru* game: "Sure, if we wrote Cannubi across the label, we could charge twice as much for it. Yes, of course, each vineyard has its own special personality; Cannubi is ripe and generous and elegant, while our Rocche in La Morra is more austere and firm. But we believe these qualities complement each other, that the grapes should be married together rather than segregated. This makes for a more interesting and harmonious wine; it also helps make a more consistent level of quality, since a particular vineyard may fare better or worse in a given year. Anyway, bottling single-vineyard wines is not something typical of this region—unless, of course, all you *have* is one vineyard."

He has adamantly declined to play the *barrique* game as well or do much of anything else to cater to the international palate or the international market. He doesn't travel (he rarely goes outside the town of Barolo, much less abroad), usually passes on group tastings and other promotional activities, and doesn't have a Web site. He does have a telephone, installed in 1989 mostly so Maria-Teresa could chat with her high school girlfriends. But when people call hoping to order some wine, he usually replies, "Ah, yes. Come on over and we'll talk about it."

Some of their wine is exported and distributed commercially but much of it, nearly half, is sold directly to people coming to purchase it at the cantina. (And sell it they do, all of it, though if they had twice as much they would surely sell that too.) Bartolo likes to do business the old-fashioned way; he prefers to meet his would-be subjects face-to-face and seems to enjoy visitors, though nonstop audiences can be a bit tiring for him. This king is an accessible one and the door to his throne room is always open—once, that is, you've made it past Fearless Franca.

Life has become much more complicated, the pace has quickened. The world has changed; wine, the local landscape, the international marketplace—everything goes spinning and spinning around. But Bartolo sits calmly at the center of it all in his horseless metal chariot, doing what he does, what he has always done, as his father did before him, and his daughter will do after. "If tastes change," I imagine him saying, "let them; eventually they will come back around again—and, in the meantime, we manage to sell all our wine anyway. If technology advances, fine; maybe we will even incorporate a thing or two, but we won't let it change our course. We are winemakers, *Barolisti,* not fashion designers. I learned from my father, rest his soul, and should he ever have a chance to taste my wine (I like to think that's where the 'angels' share' goes), I want him to recognize right away that it is Barolo—and enjoy it!"

MUCH HAS CHANGED here in the past generation or so. Many old-timers remember when it was hard to get by, when people survived on a diet of polenta, potatoes, and the occasional cat.

"The occasional *what?*" I asked when I first heard of this one evening in Renza's.

"You know, *gatto*—cat. Hey," the speaker continued, noticing my look of surprise and disgust, "they're domestic animals after all, just like chickens or pigs. And, unlike squirrels or crows, you *know* what cats eat because their owners feed them. When there wasn't enough for the people to eat, there wasn't enough for the cats to eat either, so it worked out for all concerned."

"Hmm," I said, as if the chef in me had stumbled across an exotic new ingredient. "I see your point. So what is cat like?"

"Oh, it's not bad, not bad at all, depending of course on who's doing the cooking. It's kind of like chicken but all dark meat. Actually, it can—I mean, it *could*—be quite tasty, especially if you're hungry. You have to bury them under the snow for at least a day or two to help soften up the meat and remove the gamey flavor. Longer if you want."

"Ah, so cat was primarily a winter dish?"

"Yes, that's right. *Sotto la neve* for a couple days and you're all set. Last time I had it was with him [indicating his silent companion who nods in agreement] and it was good too; *Gatto alla cacciatora* braised in wine and tomato and rosemary with some onions and carrots and celery, and lots of pepper. A dish like that along with some good creamy polenta and a *pintun* of Barbera can really chase away the coldness outside and the hunger within. That was a long time ago, of course, but it was so good I can still remember it."

(I'll bet it was too. I would try it myself if I ever had the chance, but no one around here eats cat anymore—or at least no one admits to it.)

Then the other guy jumped in.

"Hey, did you know cats have seven lives?"

"Yeah, they say that in America too," I reply sarcastically. "Except that American cats are even more fortunate: they get nine."

"No, but it's really true—I've seen it. You know that old couple who live in the run-down farmhouse up on the hill?"

"You mean the one where they blast that siren from time to time?"

"Yeah, that's it. The husband is hard of hearing, so when he's outside working and the wife needs to get his attention, she sounds the air-raid siren. You can hear that thing all the way to Alba! They're both old and not quite right in the head, especially him. Have you ever noticed how they have lots of cats around there? Well, they're not just pets, know what I mean? Those *anziani* probably don't realize the Great War is over and you don't *have* to eat cat anymore. Or maybe they just *like* it." At that, the two men, along with Renza and a couple of other old-timers within earshot, emitted a collective conspiratorial chuckle.

"Anyway," he went on, "they had this big gray tiger cat. One year it was a really cold winter with lots of snow and the old guy figured Gran Grigio's time had come. He went out, got the cat, and gave it a good knock on the head. Then, when it didn't move anymore, he stuck it under the snow behind the house. Afterward the old man forgot about Gran Grigio, just like a dog forgets about all the bones he buried. Now, the old fellow wasn't very strong and the cat wasn't dead, not really, just stunned, but it froze there underneath the snow. Eventually, spring came, the snow gradually melted, and the cat woke up. He came back from the dead! He didn't really know what happened—probably figured he just woke up from a long nap in the sun—and returned to the house to get something to eat like always. The old couple didn't think anything of it either; they didn't realize he'd been gone for months and, even if they had, surely wouldn't have remembered why. Everything returned to normal. So you see? It's really true that cats have seven lives."

"Oh, come on," I said. "You're telling me if I went up there today I would see this big gray cat?"

"Well, no, you wouldn't." He lowered his voice respectfully. "The story is true, I swear it. But you have to take my word for it: Gran Grigio vanished the following winter. Guess he had already had six lives before that!"

Many old-timers remember when they thought themselves lucky just to get by, but get by they did. They were farmers then and many remain farmers today, but things are different now. Like the mythical Beverly Hillbillies, many former peasants struck oil in the vineyards (wine) and found gold (truffles) in the woods; many now enjoy a new affluence, even though most, just beneath the gentrified exterior, retain some of the simple rusticity of their farmer forebears.

Today, Piemonte is a well-to-do region. People work hard (though admittedly less hard than they did a generation ago), but it usually pays off, and many have the substantial girth to show for their efforts. "Anyone who wants to work can find it here," they say proudly, and lots of people—Moroccans, Macedonians, Albanians, and Romanians chief among them—come here, much as Italians from the South have done for generations, to do just that.

This is an area in transition.

Many of the houses that dot the countryside are quite large (though their size can be deceptive as many contain cantinas and haylofts in addition to bedrooms and living rooms) and continue to grow as families bear fruit and add on new wings to accommodate their expanding clan and/or growing business. Construction cranes are almost as plentiful here in the Langhe as they are on the horizon of Manhattan, and buildings—residential houses, commercial *capannoni,* or new cantinas—continue to crop up all over, creating something resembling suburban sprawl.

Typically, houses have cast-iron balconies from which laundry is hung, terra-cotta tiled roofs, and a central staircase rising up through a ground floor that once stabled cows and oxen. Older houses built before the advent of indoor plumbing often have a small rectangular addition protruding from the second floor with a waste pipe coming out of the bottom, which, in its time, was an upgrade of the original detached outhouse. Most of the old fireplaces were removed when steam heat became available, but many still have wood-burning stoves. Newer houses have multiple porches and terraces—some, in fact, look like palatial villas—and many have well-tended grassy lawns around them, a rarity in the rest of Italy. Many houses, old and new alike, have fancy electronic gates and satellite dishes; it is not uncommon to see BMWs, Alfas, or Mercedes parked in the driveways and fancy fur coats in the winter.

It is hard, however, to find a nice original farmhouse (the few older ones still standing date from the 1920s and '30s, about the same age as my Manhattan apartment building). It seems that once people accumulated enough money—which almost everyone did sooner or later—they built a whole new house right next door, leaving the old (and perhaps more aesthetically charming) one to succumb to the elements and eventually crumble to the ground. When it does, the old bricks and roof tiles are collected to resell or use to lend an Old World patina to some new construction.

Most Langaroli, while not outwardly rich, live quite comfortable lives. And much of this success was fueled by wine. Looking around at hill after hill of well-tended vines, you might think that it was always like this. But it wasn't.

A mere generation or two ago, most people who grew grapes struggled relentlessly to keep afloat. Grapes were a commodity much like apples or hay, and wine little more than a beverage; even the very best Barolos could be had for a pittance—if, that is, they could be found outside the zone at all. The few wines to make it to the U.S. market were all but lost in the shadow of fine French Bordeaux and available for next to nothing.

In the postwar boom of the 1960s this situation slowly began to change, gathered speed through the seventies and positively exploded in the high-flying late eighties and early nineties. And, like phylloxera, a lot of this had to do with America. The U.S. economy was strong, a greatly enlarged, upwardly mobile, eager (and extremely young) middle class had money to burn, and many choose to burn it in the new and intoxicating pursuit of wine connoisseurship—drinking, that is.

It was during this time that one of the world's most popular and influential wine magazines, the *Wine Spectator*, was born, wine schools and clubs were formed, and liquor stores and restaurants expanded their offerings to entice this rapidly growing group of new and affluent *amateurs*. The meteoric rise and coming-of-age of the California wine industry (which had been severely crippled but not quite extinguished decades earlier by Prohibition) occurred in tandem with these other domestic developments, and new and promising areas such as Long Island, Oregon, and Washington State jumped onto the merry wine wagon.[1]

Suddenly, the modest output of small, previously obscure Piemontese wineries were snapped up by American importers who quickly sold out whatever they got. Over in Barbaresco, Angelo Gaja emerged as an international player; both a phenomenal winemaker and a marketing genius, he demanded extraordinary world-class prices for his wines—and got them. Many others followed suit—even some who had never before made wine or grown grapes jumped into the fray—and the cost of Barolo and Barbaresco, now often garbed in fancy bottles with chic labels, skyrocketed, as did the wholesale cost of grapes and vineyard land within the designated zones.

This radically changed the demographic fabric of the region. Aside from significantly improving the economic well-being of farmer and winemaker, it induced their children not only to remain in the family business but to go to agriculture or enology school, sometimes even in faraway places like UC–Davis, where they learned the latest technology and marketing strategies. Few, however, felt compelled to go back to working the land as their fathers had, and it became necessary to hire people from outside—low-skilled, low-paid immigrants—to perform the basic vineyard tasks. Today, it's becoming increasingly rare to find native Piemontese in the vineyards of Barolo and a winemaker working among his vines is becoming almost as unusual as a New York celebrity chef actually cooking in his kitchen.

The countryside changed too. People looking back as little as fifteen or twenty years ago saw a radically different landscape. There was a more diversified patchwork. There were vines to be sure, but there were just as many groves of hazelnut and walnut trees, stands of dense forest where truffles grew and wild animals lived, and yellow patches of wheat. ("You can do without wine if you have to, but you can't live without bread," said an old-timer who remembers when there wasn't enough to eat.) Hazelnuts, particularly the indigenous Tonda Gentile, remain a part of Piemontese culture and are still used in traditional sweets such as *giandujotti*, Torta di Nocciole, and *torrone* (nougat), but their cultivation is shrinking, or at least becoming more of a strictly commercial enterprise relegated to areas where vines won't grow. Like many other industries, the hazelnut business is becoming consolidated; it is more cost effective to harvest large tracts with expensive machinery than small individual groves by hand. Fabrizio ripped out the family hazelnut trees a few years ago, replanting a vegetable garden in its place: "It was just too much work to collect them all," he said. "Hard work too, bent over for days on end picking all those nuts up off the ground. And for what? With the price so low, it's cheaper to buy them than to sell them." Ivana remembers having to help harvest the hazelnuts as a kid and was happy to see the trees go.[2]

The once-ubiquitous Madernassa pear, a cooking pear traditionally poached in Barolo wine or used in Cugná, is also becoming increasingly rare.

The Langhe is on the verge of becoming a monoculture. Never before has there been as much territory under vine as there is today. And it is still increasing. This has had many benefits, to be sure, but it has also created some problems, both ecologic and economic. One is erosion. Vines have extraordinarily deep roots (especially relative to the small size of the plant), but, precisely for this reason, they don't anchor much of the topsoil. More and more forests of trees and shrubs are being cleared and converted to viticulture, thus creating vast areas of loose, tilled soil with little to hold it together in heavy rains. Many winemakers are digging big drainage ditches or installing extensive drain and tube systems in their vineyards to carry away excess water with minimal erosion.

On the economic side, the increase of land under vine (along with more sophisticated agricultural techniques and equipment) has increased overall production, further contributing to the glut of wine on the international market. This surplus has had a reverse effect of lowering the price and overall profitability of wine while increasing operating costs. It has also induced some winemakers to spend more on marketing and promotion to help them survive in an increasingly competitive industry.

This downward cycle was in the making for some time but has recently been accelerated by outside events. With the bursting of the dot-com bubble, the fall of the World Trade Towers, the rapid rise of the U.S. deficit, and the plunging value of the dollar, the American market for Barolo has dried up, leaving many producers with lots of unsold wine in their cellars and lots of debt incurred to increase production, renovate wineries, or purchase new land.

A cloud hangs over the industry; will it burst open or dissipate? And when? Whatever happens, the immediate future for the Piemontese winemaker does appear nearly as light or bright as the late-spring sun that, as I write, is shining down on their vines and ripening their grapes.

· RECIPE #15 ·

TORTA DI NOCCIOLE

Classic Hazelnut Cake

INGREDIENTS

2 heaping cups (approximately ¾ pound) hazelnuts, preferably Tonda Gentile (see note below)

1½ sticks (6 ounces) butter, softened (plus 1 tablespoon additional butter for greasing the pan)

1¼ cups (10 ounces) sugar

2 tablespoons extra-virgin olive oil (or substitute a good-quality hazelnut oil)

2 eggs and 2 egg yolks

1 cup sifted "Tipo OO" flour (or substitute pastry flour)

1 envelope Pane d'Angeli yeast (or substitute 1½ tablespoons baking powder plus ½ teaspoon vanilla extract)

1. Preheat oven to 350 degrees. Arrange the hazelnuts on a baking sheet, and toast, shaking the pan occasionally, for about 10 minutes. Using your bare hands or a kitchen towel, rub the toasted nuts gently, removing only the skins that fall off naturally. Let cool.

2. Place the toasted nuts in the bowl of a food processor fitted with the chopping blade and pulse to chop finely and evenly.

3. Place the butter and sugar in the mixing bowl of an electric mixer fitted with the paddle attachment and cream together. Mix in the olive oil, then the eggs and egg yolks, one by one, mixing well each time to thoroughly incorporate.

4. Fold the chopped hazelnuts into the batter, and mix thoroughly. Combine the flour and the yeast and add to the bowl little by little, whisking to make sure it is well blended.

5. Butter the sides and bottom of an 8-inch spring-form pan. Pour the batter into the pan (it should come a little more than halfway up the sides), place the pan on a baking sheet, and bake in the preheated oven for 40 minutes, or until a toothpick inserted into the middle comes out clean.

6. Turn off the oven and let the cake sit inside, without opening the door, as it cools down (about 15 minutes).

7. Remove from the pan and let cool on a pastry rack for another 15 to 20 minutes.

YIELD: SERVES 6 TO 8.

———

NOTE: The indigenous variety of Piemontese hazelnut, Tonda Gentile, is smaller than most others and prized for its aroma and full, rich flavor. If you cannot get this variety, use another. Just make sure the hazelnuts are fresh (the longer nuts sit around, the more they are apt to lose their flavor and turn rancid).

This cake is delicious but a bit dry. It begs for a glass of sweet wine to accompany it—Moscato d'Asti or Brachetto d'Aqui, perhaps. Or better yet, a Zabaione made with Moscato or Barolo.

· RECIPE #16 ·

ZABAIONE AL BAROLO

———

ZABAIONE (SOMETIMES ALSO spelled *zabaglione*) is a custardlike dessert served throughout Italy. Traditionally, it is made with sweet

Marsala wine, though here in Piemonte it is just as often prepared with Moscato or Barolo. (I actually prefer the latter as it is less sweet and has a lovely pink color.)

The basic recipe is always the same: egg yolks, sugar, and wine are combined and whisked over a double boiler until thick. The amount of sugar must be adjusted depending on the type of wine used: if using a sweet wine such as Marsala or Moscato, you would need to use less sugar, while a dry wine like Barolo would require slightly more.

Usually, Zabaione is served hot, but, if it is allowed to cool, you can fold whipped cream into it and serve it with summer fruits and berries.

INGREDIENTS

4 extra-large egg yolks
4 scant tablespoons sugar

4 generous tablespoons Barolo
(or other dry red wine)

1. Put the ingredients in a heavy-gauge bowl, preferably of copper or stainless steel (*see note below*).

2. Place the bowl over a pot of boiling water to create a double boiler and whisk constantly until thick about 10 minutes. When done, it should have the consistency of thick custard.

YIELD: SERVES 4 AS A SNACK OR AS AN ACCOMPANIMENT TO TORTA DI NOCCIOLE (PRECEDING RECIPE) OR PEARS POACHED IN BAROLO (FOLLOWING RECIPE).

NOTE: Copper or stainless steel are preferred, as the egg yolks may react unfavorably to aluminum or other metal. A heavy-gauge bowl is important too; if the bowl is thin, the eggs may cook too quickly and/or stick to the bottom and curdle. There are fancy copper bowls with a long handle specially made for Zabaione. In the absence of one of these, you can simulate the thickness of copper by putting two or three stainless bowls together. You may find it necessary to adjust the heat of the double boiler and/or take the bowl of Zabaione off the heat from time to time while it's cooking to prevent scrambling.

· RECIPE #17 ·

PEARS POACHED IN BAROLO

INGREDIENTS

1 bottle (750 milliliters; about 3 cups) Barolo, nebbiolo, or other red wine
½ cup sugar
4 pieces star anise
2 cloves
1 cinnamon stick
6 to 8 pears, preferably Madernassa or Martin Sec, or substitute another small, firm cooking pear such as Seckel

1. Put the wine, sugar, and spices in a pot and heat, stirring, just enough to dissolve the sugar.

2. Peel and core the pears (*see note below*) and place in the pot. There should be enough wine to cover them.

3. Raise the heat to medium and let cook on a gentle simmer for about 20 minutes or until a knife goes into the pear without resistance.

4. Turn off the heat and let the pears stand in the liquid for at least 1 hour. (May be made ahead to this point and stored in the refrigerator in a plastic container with the wine covering the pears.)

5. Remove the spices and discard. Remove the pears and, in a heavy-bottomed saucepan, reduce the wine until it has the consistency of syrup.

6. Slice or stuff the pears (*see note below*) and drizzle with the Barolo syrup.

YIELD: SERVES 4 TO 8, DEPENDING ON HOW THEY ARE BEING SERVED

NOTE: Peeling and coring is a matter of opinion and, ultimately, how the finished pears will be used. Many Piemontese cooks don't peel the pears at all, but the skin turns wrinkly after it's cooked and may be a bit tough. I generally prefer to peel them, which allows the color and flavor of the wine to permeate the pears. Try coring the pears from the bottom up, leaving the top with the stem intact, and then stuff them with a mixture of Gorgonzola cheese lightened with mascarpone or sour cream.

The slices can be served over vanilla ice cream or Zabaione.

21 • WAR AND PEACE

BACK IN LE MUNIE the vines continue their promiscuous growth, blissfully ignorant of events unfolding in the man-made world.

One morning in early June, Fabrizio comes to pick me up. We have a quick espresso, get in the car, and head up the hill in silence. He seems preoccupied. After a few moments he says gravely, "We're heading into war."

"What?" I asked, alarmed.

"You'll see."

Shortly after 9/11, America had invaded Afghanistan and easily tumbled the Taliban. Now there was talk of going after Saddam Hussein in Iraq as well, next on the Axis of Terror list, and America was actively trying to create a coalition of allies and obtain the approval of the United Nations. In Italy, this sword-rattling was extremely controversial. President Bush was viewed as a gunslinging cowboy, a Texas oil mogul who wanted to take over Iraqi oil fields and get back at Hussein on behalf of his father, George senior. As an American, I was frequently put in the uncomfortable position of having to defend his actions, even though I didn't completely agree with them (I was usually able to wiggle out of it by raising the subject of Prime Minister Silvio Berlusconi, himself something of a spaghetti-western cowboy).

Throughout Italy, multicolored striped flags with the bold letters PACE hung from balconies and windows everywhere (the churches that distributed them had completely run out of stock), and there was a collective feeling of tension in the air.

This whole situation made me uncomfortable and edgy. So when Fabrizio first said we were going to war, I took it literally. I imagined I was being drafted into some local branch of the Alpini and that we were on our way up to the piazza in Monforte to be fitted for a beige uniform of below-the-knee pants, high wool socks, and a little cap with a feather, and, amid banners and tears, shipped off to the deserts of the Middle East.

When, however, we took the familiar turn off the road to Le Munie, I breathed a little easier—at least at first.

It had been well over a week since we had last been here and things had got out of control; the vines were going crazy. Some time before, we had pulled the third set of nylon cords near the top of the posts. At that point the tips of the vines had barely touched them. Now, however, they were shooting out in all directions, some within the two upper cords but most outside, well beyond the height of the posts. Many of the branches were more than six feet tall, and they were getting thick. The once tender young shoots had grown into rambunctious adolescent vines, and these in turn had grown into strong, woody warriors. What's more, these unruly combatants were actively sending out reconnaissance parties, staking out new territory, and holding their ground by attaching strong tendril trusses that grabbed on to anything in reach and held on to the death.

From the top of the gravel road, we surveyed the scene and planned our strategy; our opponent was formidable, and victory was not to be taken for granted.

"Ready?" Fabrizio asks. "The battle is about to begin."

Our objective is to get the branches to go as straight as possible up from the base of the *capo a frutto* in between the two upper cords and secure them there. This is important, especially during the critical months ahead leading up to the harvest. If the branches are overlapping, the grape bunches will be piled one on top of the other and will not ripen evenly. The leaves need maximum exposure to the sun, and the grape clusters, growing ever larger and heavier, need room to breathe in order to mature fully and without rot. However difficult straightening out the mess may be now, it will only get harder as time passes and further complicate later operations. It's now or never.

Get ready, set, *charge!* We plunge into the thick green trenches with a frenzied rush of adrenaline. Grab a branch, pull it free, force it in between the cords, and lop off the excess. Once a group of vines have

been thus ordered, we must secure the perimeter by tying the two nylon cords together, herding them in like prisoners of war.

This is furious hand-to-hand combat. Casualties are heavy on both
sides: yank on a branch, and it suddenly pops out, smacking you in the
face (one for them); tough green tendrils refuse to let go and must be
savagely cut free with the clippers (one for us). The bloody skirmish
rages for days but in the end we prevail; we have imposed some order
on the vineyard, at least for the time being.

At dusk of the third day it is over. We leave the battlefield littered
with fallen leaves and severed branches, bruised and battered but tri-
umphant, and head to the bar to nurse aching muscles, scratched faces,
and blistered hands with our hard-won booty, a celebratory glass of
wine.

After the victorious Battle of Le Munie, Fabrizio and I return periodically to keep the peace, administer our conquered territory, and
stifle any attempts at uprising. There are three main tasks at this point;
taja i gich, pulidè, and *dare l'acqua.*

Taja i gich is the Piemontese term for trimming the branch tips, the
part of the vine (now pushing eight feet tall) that extends beyond the
height of the top wire. Traditionally, they were draped over one another and twisted over the upper wire, creating a long bulge along the
top of the row. Most people, however, feel that trimming the tips is a
good idea; it limits growth and once again focuses energy down to the
lower fruit-bearing portion of the vine. It also makes it much easier to
take down the dried vines during first pruning the following winter.

There is a machine (a tractor attachment, actually) that can do this,
but in most small *aziende agricole* it is still done by hand. It's easy; just
make sure the branches are more or less straight, measure a good hand
span above the top wire, and cut away. It's like hedge trimming, after
which the clipped rows of cut vines look like well-treated detainees
with jagged buzz cuts lining up for morning roll call.

Pulidé means to clean *(pulire)* the vines. The still-green grape clusters must be untangled so they hang straight down, and any leaves and
small branches in the way must be removed. This allows the grapes to
grow unencumbered and mature properly; it also makes harvesting

easier and helps prevent rot. Most of what comes off are the *femminile*, little shoots popping up between branch and leaf or *doppie*, sizable double branches shooting off the main ones. Any extraneous, unnecessary growth around the cluster is removed, and the interior of the vine is thinned out as much as possible, allowing light and air to circulate. It is important, however, to leave an outer layer of leaves on both the front and back intact so as to provide protective shade for the grapes, especially in a hot year like this one. If this pruning is done correctly, the grapes should be able to happily soak up the heat without getting burned, like bathers lounging on the beach under the protective shade of a nice big umbrella.

Dare l'acqua is a sort of euphemism. Literally it means to "give water," but this water is actually mixed with copper sulfate and is the most common chemical treatment.[1] The use of copper and sulfur to prevent mold on grapes and preserve the wine goes back at least to the time of phylloxera (American rootstocks were resistant to the deadly aphid but more prone to mildew than vinifera) and perhaps even back to Roman times.[2]

Here in the vineyards of Barolo, you can still find many old circular cement vats, their insides dyed blue by copper. They are about three feet in diameter by three feet high and heavy as hell. Originally they were used to mix the chemicals with water (often from a nearby well) just prior to application, but they are now little more than trash receptacles. Today the mixing usually takes place in a big plastic tub right on the back of a tractor bed in the cantina.

There are several ways to apply the *acqua*. The most common and efficient (or at least time saving) is a tractor attachment that dispenses the liquid by blowing it with a fan throughout the vines on either side of the row. This requires a modernized vineyard with rows wide enough apart for a tractor to pass as well as yet another piece of expensive machinery. Some large properties use helicopters to spray from the air, which is fast but hardly precise or judicious. The old way was for a farmer to walk up and down the rows with a rectangular can strapped to his back containing the copper-sulfate powder and blowing it out through a nozzle with a hand-operated pump.

Fabrizio and I use the next oldest method. (He doesn't have the fancy tractor attachment; in fact, he has only just recently acquired a small secondhand Pasquali tractor and many of the rows in Le Munie are too narrow anyway.) A large square plastic vat containing the pre-mixed *acqua* is mounted on the tractor trailer. Also mounted on the trailer is a wooden spool with a long rubber hose. One end of the hose is attached to the vat, and at the other end is a long spray nozzle. The tractor trailer is backed into the central aisle of the vineyard just below the gravel road and a brick is placed under each of the wheels to prevent it from rolling down the hill. We then pull the hose down to the bottom of the vineyard plus a generous allotment to compensate for the width on either side.

When he's ready to begin, Fabrizio, down at the bottom, gives me a signal; I turn on the tractor and put the pump into gear. I then hurry back down to the bottom as Fabrizio begins to spray the vines.

The goal is to get an even covering on all the leafy sections of the vine, being careful not to miss any spots but also avoiding total inundation. In order to do this, you have to move continuously down the row, not too fast but not too slow either, moving the face of the nozzle up and down the vines in a gentle but controlled figure eight.

Dare l'acqua is, at least in this primitive way, a two-man operation. My job is to *"tiré la goma"* (to pull the rubber hose). As Fabrizio heads down a row, I need to feed slack, and when he comes back again, I need to pick it up, making sure not to let the hose get tangled up in the process.

We work our way up and down the rows, from the bottom to the top of the vineyard. Before long, Fabrizio is completely drenched with *prodotto,* and I'm drenched too, a mixture of sweat and liquid from the wet hose. The last two rows are mine to spray.

This is my least favorite of vineyard tasks; the chemicals are smelly, the tractor engine is noisy, and I can't seem to find the right pace or rhythm moving down the row or the right graceful motion of the wrist with the spray nozzle. But giving *acqua* is necessary to protect the grapes. Usually, it must be applied every seven to ten days during this critical growth period and especially after it rains. But this year it doesn't rain. This year the special *acqua* is applied less often than usual—hardly at all, in fact—for this is to be an unusual summer, quite unlike anything anyone around here can remember. This is to be the summer from hell.

PART III

SUMMER
INTERLUDE

22 · INFERNO

EARLY IN JUNE a stifling heat descends on the Langhe—on all of Western Europe, in fact—an unusually intense heat that doesn't let up for months. The dancing *nebbia* of the preceding winter has taken off for a long vacation only to be replaced with a thick, perpetual *afa* (haze) that completely obscures the snow-capped mountains.

Castiglione retreats indoors, its scorched inhabitants in a state of sweaty shock.

"It's never like this here, really!" they say.

"Yeah, well, it's sure like this now!" I think to myself as I wipe off the beads of perspiration dripping down my face.

This is my first summer here, so I have no basis for comparison. But I am surprised; I didn't expect it to be so hot and humid, not here in the foothills of the Alps, and especially not in June. I am not alone.

In France, people are literally dying from the heat, which creates a public scandal.[1] Yet no emergency cooling stations are set up—as they often are during extreme hot spells in the States—where people with no other recourse can go to escape the danger. Apparently, even the socialistically inclined Europeans were not prepared for this. There are frequent warnings for old people and those with respiratory conditions to stay inside. But that is hardly a solution. It doesn't cool off in the evenings and the thick stone walls, instead of offering an insulated buffer from the heat, seem to retain it. Practically no one has air conditioning—most don't even have fans.

Out of desperation, I go into town and buy a little oscillating fan. When I get home, I rip it out of the box, plug it in, and watch it rotate back and forth with a sense of joy and amazement as if I had just discovered fire. That night I proudly move it into the bedroom and carefully position it at the foot of the bed. But Ivana can't stand it. Like many raised in the country, she is not accustomed to an artificial breeze:

"Non mi piace," she says adamantly. "It is unhealthy. Turn it off."

Instead, she prefers to keep the windows open, letting in the stifling, stagnant air and a never-ending onslaught of mosquitoes and nasty little black bugs, both of which seem thoroughly thrilled at having a foreign delicacy to gorge themselves on. Is my flesh tenderer? I wonder. My blood tastier? Or is it just a question of easy access? Ivana (who doesn't seem to be bothered by the insects) is soundly asleep covered up in a sheet like a mummy while I am splayed moist and naked on top of it. I play a vicious game in the dark, trying to squash the bugs before they land, puncture my skin, and suck my blood. My adversaries, however, are not easily vanquished. The mosquitoes are smart; they work in teams, one buzzing around my head as a decoy while the other attacks my leg, but the black ones are even worse—they make no sound at all. Sooner or later I give up and retreat into the lighted living room, position myself in front of the fan, and keep watch against the bugs. By mid-June my legs are covered with bloody welts.

During the day, there's nothing to do but move slowly, keep out of the sun, and take periodic cold showers. People do their shopping early in the morning and then rush back inside. Sometimes, stricken with cabin fever, I venture out after five for a bike ride; at least I can get a breeze coasting downhill. But, for the most part, it is best to stay inside. And there's really no reason not to. Work in the vineyards has come to a complete stop and the earth has turned to dust. At this point all you can do is watch the grapes bake in the sun and hope for the best.

And it goes on like this week after week without respite.

Often, in the late afternoons there are rumblings and ominous black and blue clouds off in the distance. Nature flexes its muscles. A dreaded *temporale,* the feared fast-moving summer storm, hovers off in the distance, and the *casalinghe* (housewives) rush into action, pulling in the laundry, closing the windows, and securing the shutters. It is dreaded most of all by the farmers because the collision of cold moisture-filled clouds with hot air can create dangerous hail, like the storm that did such extensive damage in 2002. Almost a year later, you can still see the swaths it cut through the vineyards. At its worst the hail is golf-ball-sized and rains down dry like a shower of icy rocks, denting car bodies and knocking out power lines—imagine what it can do to grapevines! Even pea-sized hail can poke holes in the grapes, decimate foliage, and cauterize the stems, cutting off circulation of essential nutrients to the clusters; the whole plant just shuts down.

You never know when the *grandine* is going to strike or exactly where; one vineyard might be totally devastated while another right next door is spared. A whole year of hard work can be wiped out in a matter of minutes, and there's nothing you can do about it.[2]

People want the rain but without the hail. These conditions, however, are perfect; it is not so much a question of *if* it will hail but when, where, and how much. Cannons are fired off in some communes "Rainmaker"-style to break up the storm clouds, despite doubts as to whether it really works or not. And some people perform an old local rite of throwing the sanctified olive branches from the previous spring's Palm Sunday Mass into the purple shadows of an oncoming storm, thus invoking divine assistance to prevent hail. But this year these precautions are unnecessary: there just isn't enough moisture to create rain, much less hail, and the threatening clouds usually dissipate with nothing more than a light misting. It does hail once or twice, but only small lentil-sized pellets, only for a few minutes, and always mixed with water— nothing to cause any real damage, but nothing to offer much relief either. The scant wetness is quickly sucked up by the parched soil and sizzling asphalt, and within the hour everything is dry again.

Weeks pass and the heat continues. On the front page of the newspapers are horror stories from France, photos of sunburnt fields and talk of rationing water. And no one will venture to guess when the situation will change. Fruit and vegetable growers are extremely hard hit; prices are expected to rise and stay there for some time.

This is an agricultural community, and everyone's livelihood hangs, to a greater or lesser degree, on the weather. You can feel the tension in the hot, sticky air. Farmers strain their necks looking up at the glaring sun, down at their dusty vineyards and wilting vines, and back up to the sky. For the most part they remain stoically silent but their concern is obvious. "Last year endless rain and hail; this year endless drought. *Che brutta la vita del contadino!*" I imagine them saying.

But whatever they might be saying or thinking, the Powers That Be don't seem to care—the summer of 2003 rages on relentlessly.

23 · PARTY ON!

SUMMERTIME IN ITALY is *festa* time, and it would take much more than the heat wave of the century to stifle the irrepressible Italian urge to party.

Throughout the Langhe—throughout the entire country, in fact—an unbroken chain of festivals stretches from one end of the summer to the other. Sometimes the *festa* is in honor of a local patron saint, sometimes in celebration of a local foodstuff, and sometimes there is no ostensible excuse whatsoever. It often happens that there is more than one *festa* on a given day and you have to choose, or, if you are ambitious, go *festa*-hopping from one to the other. The action reaches its peak around *Ferragosto* (August 15), an unofficial national holiday when everything comes to a complete halt, after which summer tapers off into the twilight of autumn like the trailing sparks of a big, colorful *fuoco d'artificio*. There are festivals throughout the rest of the year, to be sure, but they are fewer and farther between and lack the pervasive party atmosphere of the nonstop summer *festa* season.

One of my first experiences of the Italian *festa* phenomenon was in Castiglione at a feast in honor of the village patron saint, Sant'Anna. It took place under the *peso pubblico* (public scale) at the foot of town. There was a barbecue, keg beer, and local red wine served in plastic cups. And there was a local hard-rock band belting out high-decibel music to a largely older, unenthusiastic audience of people from the village. Things livened up a bit, however, when the band shifted into seventies disco favorites like "Y.M.C.A." As if on cue, nearly everyone lined up and moved their arms in sync like some multigenerational cheerleading team rehearsing their routine. As I was soon to learn, group dances are extremely popular in Italy.

On August 11, during the Feast of San Lorenzo, everybody goes outside and stares up into the night sky to watch for shooting stars; if you see one and make a wish, it will come true (unfortunately, during the heat wave of 2003 you couldn't see anything through the thick, steamy haze). And at the tail end of summer is the Feast of the Madonna, when people throughout the Langhe and Roero light bonfires called *falò* in her honor. It is magical indeed to go out into the desolate countryside on this night and watch the distant fires flickering like relay points through the darkness, while you inhale the pungent smell of wood smoke splattered liberally on the late-summer air like perfume on a dowager princess. The Madonna, one gets the feeling, is pleased.

Hand in hand with many festivals are other activities such as crafts markets, farm-machinery exhibits, fashion shows, and tournaments of an odd, indigenous street game without the street called *palla a pugno*.

At central locations in many small villages throughout the Langhe, observant visitors may notice what resembles a derelict wooden storm door dangling at a weird angle on the wall of an old building. (I had certainly noticed the one on a lower portion of the Castiglione castle, but assumed it was just something someone had not gotten around to taking down.) Though it sits there untouched for most of the year, during the summer the old door is raised up on its rusty hinges and propped against the wall at a 30-degree angle, turning the surrounding area into an impromptu playing field.

Palla a pugno (fist ball), which later also became known as *pallone elastico* when balls of rubber replaced the original leather ones, likely began in the courtyards of farmhouses and reached its peak a generation or two ago when many of the fancier public "courts"—basically just large, centrally located flat areas with a wall on one side—were established. Every village had its team, which competed against neighboring villages, and outstanding players became local celebrities.

These days, with the widespread popularity of soccer and many more sophisticated diversions to choose from, the game has dwindled to something of a cultural relic hauled out during summer festivals. I got to see it played on several occasions, and while I can't say I was able to understand how the game works, at least I was glad to finally figure out what the weird door was for.

Any respectable village hosts at least one or two *feste* during the summer, but few can rival the Festa della Bussia.

The Bussia, though geographically closer to Castiglione, is officially part of the commune of Monforte. (In Italy, as in rural areas of America, the boundaries of a town, or *comune,* are much larger than the limits of the village, or *paese,* at its center).

Monforte *paese* is a beautiful village. It is reputed to have been originally founded by the Cathars, a mysterious Christian sect, but because of the group's strict vows of celibacy and their natural tendency to keep a low profile due to relentless persecution by the Catholic Church (which considered them heretical), little evidence of their presence here remains.

The town is divided into two parts, Monforte Alto (or Vecchio) and Monforte Basso. The lower part is a bit newer, and here, just off the main piazza, is the *municipio* as well as the ornate Chiesa della Madonna di Neve (Church of the Madonna of the Snow), where Ivana was married. The *vecchio* part is located up a steep hill of twisting cobblestone lanes flanked with narrow houses, and at the top of the hill sits an old church and bell tower. From this plateau, terraced stone and grass cascade down one side forming a natural open-air amphitheater, where, just outside the wall of the *castello,* concerts are held in warm weather.

When I first visited Monforte about fifteen years ago, this old part of town was largely abandoned. Now the decrepit old houses have been reclaimed, restored, and reoccupied and are more beautiful than ever. But most of the local characters—characters like *Barba Rusa* (Red Beard), who practically lived on the main piazza at the center of town, terrorizing the tourists, and Camiot, who was often seen driving around in his dirty beat-up car filled with goats and dogs, going from farmhouse to farmhouse to sell his cheese (he used to call on Ivana's mother)—have passed on for good.[1]

Like many towns, Monforte has a number of satellite communities called *frazioni* or *borgate.* But the Bussia is something else, even smaller and more diffuse than a *frazione.* It is what is known as a *località,* simply a "place." Actually there are two: the Bussia Soprana and the Bussia Sottana (upper and lower).

Some say the name Bussia comes from the word *abisso,* which means abyss, depth, or precipice. In Piemontese, *bussia* refers to a piece of bronze or metal used as a guide or fulcrum. Both explanations seem to make sense. From the top of the Alba-Monforte road you look down on the vine-covered gully between the upper and lower Bussia falling off to the west like a big diagonal wedge cut out of the earth.

Going down the road to the right, the road I sailed down in darkness months before when I first went to Ivana's for dinner, is the Località Bussia Sottana, a mere collection of four or five houses (one of which is Bruna's, one Ivana's). Farther down is the pink farmhouse where Silvana, mastermind of the portentous white-truffle dinner, lived before she got married, and just beyond that the road bottoms out as it crosses the border into the town limits of Barolo.

If, on the other hand, you go up the road to the left and take the first turn off the Bussia Soprana road, you'll go up another, steeper incline at the top of which is an old church. The church (though not deconsecrated) is seldom used anymore, but its bell tower still stubbornly rings out the hours, twice in fact for anyone who might have missed it the first time. And attached to the church is a building that once housed the one-room schoolhouse where Ivana went as a child. These buildings are surrounded by a more or less level threshold, and this is where the Festa della Bussia takes place.

I had heard of the *festa* from Bruna, who always found a way of returning to Italy each summer, as if to fulfill some mandatory civic obligation. But I had little idea then of what an Italian *festa* was and even less what a *"bussia"* might be. Now, however, I was intimately familiar with the Bussia and was finally going to see the *festa* for myself. Actually, I had been drafted to help out.

This party goes on for five days in the middle of August right around *Ferragosto,* and there is no excuse for it other than to celebrate. It's like a big block party thrown by residents of the neighborhood, and almost everybody pitches in. Giulio (who has a farmhouse and cantina in the lower Bussia) is in charge of the ravioli, Franco leads the barbecue team, and Aldo (one of Bruna's two brothers and a surprisingly excellent dancer) organizes the music. Other residents man the ticket booth and the bar, where you can get bottles of wine made by Bussia producers, and draft beer. Fabrizio is currently president of the Bussia Pro Loco and responsible for overseeing the whole thing.

Notwithstanding the big buildup over the past several months, I had no idea what was in store. When we arrive at the Bussia Soprana on a sweltering Thursday evening just before eight, the whole area is cordoned off and guarded by an army of men and women in bright orange vests with the words PROTEZIONE CIVILE written on them. A fully extended cherry picker is parked off to the side, and from it, right over the center of the road, dangles a big sign with the letters B U S S I A spelled out in lightbulbs. As we approach, the barricades swing open and we are waved through. We park in a reserved area at the base of the hill and walk up the rest of the way on foot. Once we reach the top, panting lightly from the effort, the remainder of my breath is taken away by the transformation of this normally pastoral little spot.

At the edge of the plateau behind the church, a huge, high-peaked, gazebo-like dance floor, framed by arched doorways and wooden railings, has been erected. The thing resembles an immense, horseless merry-go-round, and at its rear is a raised platform on which a group of eight or nine musicians is tuning up. Along the entire length of the church building, a vast patio has been filled with long wooden tables and hundreds of white plastic chairs, the whole thing covered by a tarpaulin roof secured to a rusty metal frame. Strings of little lights dangle around the perimeter. On the other side of the church, where the card tournament will take place, are more tables and chairs, and at the far end a tarp-covered area where ribs and ravioli will be prepared. ("It has NEVER rained on the Festa della Bussia," Fabrizio says, "but you never know.")

As Ivana and I make our way across the grounds, we stop frequently to greet people, Ivana introducing me over and over again. I quickly realize that this is her extended family, people whom she has known, lived among, and celebrated with her entire life, and that for her, having me here must be kind of like bringing the boyfriend home for the holidays to meet the folks.

It takes a while, but eventually we reach the food area, which is divided into two sections, each fronted by a long low table. At the rear, open wood fires are raging. On the far side, a bed of smoking charcoal is covered by a big iron grate. On our side, a fire is burning under a metal frame that supports a gigantic pot of water, and a bit farther off is an old beat-up enamel butane stove with smaller pots of meat sauce bubbling away.

Ivana stashes her purse under the table and introduces me to our *capo*, Giulio. (Alan Tardi, American, former six-figure-salary New York City chef, reporting for duty, I think to myself as we shake hands.) They give me my instructions along with a frilly bib-apron, which I tie on over my moist tank top and shorts. "When people come, you take their yellow ticket—that's the one for the pasta (the pink one is for the *costine*)—put it on the spindle, and give them a plate of ravioli, one for each ticket. If anyone tries to cut ahead, you tell them they have to wait in line. Okay?"

"Got it!" I reply.

We all chat and joke amiably while the water continues its long, leisurely journey to a boil. Someone has scored us a bottle of wine to drink out of plastic cups. There is a pervasive festive atmosphere, even though the *festa* has not yet officially begun, and I soon figure out that this party is first and foremost for the residents of the Bussia themselves. But there *are* guests.

At a certain point we turn around and notice a long line of people stretching out along the the table, each clutching yellow tickets in their hands. Giulio puts down his wine and moves into action, sliding ravioli from a paper tray into the pot, though the water still has not yet come to a boil.

"*Due minuti, eh, due minuti. Pazienza, per piasín!*" he says to no one in particular.

When finally cooked, the ravioli are transferred into a large rectangular pan and mixed thoroughly with the meat sauce. Giulio slowly tastes—countless hungry eyes fastened upon him—and then gives the thumbs-up. With that, we dive in, spooning ravioli onto plastic plates, running back and forth like crazy passing them out to the starving masses in exchange for a wrinkled yellow slip of paper. In a flash they are gone and the whole thing starts again. Before long, we catch a rhythm and the *festa* is off and running, broken only by lag-time to let the water come back to a boil or by coworkers, friends, or residents of the Bussia, the Protezione Civile or Carabinieri, who get to cut ahead of the line, much to the dismay of the patient ticket-holding (paying) guests.

The empty sea of white plastic chairs is soon filled with a bobbing mass of humanity. People wander around with plates and wine bottles in hand, looking for a place to sit; even the two steps of the little church are occupied. At around nine it starts to get dark, the lights come on, and the band begins to play.

I have another one of those magic moments: a warm summer night; the smells of hot bodies, wood smoke, and sizzling meat mingle on the stagnant air. I am here at the Festa della Bussia *behind* the table, wearing an apron, hanging out with my new Bussia friends, drinking cup after plastic cup of wine (in some cases with the very person who made it), and trying to remember everyone's name. Lanterns twinkle in the otherwise imperceptible breeze, and now, to cap it off, live dance music comes drifting back through the thick air above the murmurs of the crowd.

Before long, however, my idyll is abruptly shattered; the music stops mid-song, and a frantic call for medical assistance is broadcast repeatedly over the loudspeaker. We continue working but news travels quickly: someone has had a heart attack right on the dance floor! Then we learn that the man has actually died. (Other rumors begin to circulate that the woman he was dancing with was *not* his wife.) Out of respect and propriety, the band will not play for the rest of the evening, which puts a bit of a damper on things. But there is a certain amount of grim humor too:

"Sure, I feel bad for him and his family," someone says, "but what a way to go! Dancing with your girlfriend at the Festa della Bussia—*una bella morte, ne?* It doesn't get much better than that. We should all be so lucky!"

This was not the first tragedy to strike the *festa* (someone died here two years ago), and this raises immediate concern among some of the organizers that people might begin to think the festival is cursed. But others try to make light of it, suggesting they create T-shirts for next year's event reading *"Festa della Bussia: Bella da Morire!"* (To Die For!). Maybe, someone suggests, they should require a health certificate to get in. Food and wine service continue throughout the evening, but, without music, the atmosphere is undeniably muted.

This year's *festa* got off to a bad start, but luckily there are still four more days to go and they pass without major incident. Midway through the second night, we get a break and Ivana pulls me back behind the food area and places my hands firmly on her round behind. *"Guarda!"* she says, indicating the Alba-Monforte road below. Cars, headlights aglow, are lined up in both directions for as far as you can see, and they're all heading here. It looks like a double string of pearls draped over the darkened hills, shimmering in the night. *"È bellissssima, no?"* "Yes, it is," I want to say, but she gives me a long kiss before I can reply.

The third night was the best of all: it was Saturday, it was *Ferragosto*, and Ciao Pais was playing. Ciao Pais is the best of the bands and, with long and strong ties to the *festa*, the clear favorite. Like the others, it is a large group of accordion, saxophone, upright bass, clarinet, drummer, and a sexy female vocalist or two; at certain points they too break into disco favorites to give the kids a chance to get involved. But what they really specialize in is traditional music for *ballo liscio*.

Dancing (especially *ballo liscio*, "smooth" couples dancing) is, along with wine and food, an indispensable part of the *festa*'s "Holy Trinity," and I have never witnessed anything quite like the scene that unfurls each night under the peaked roof of the open pavilion. When the music starts, everybody—hefty middle-aged women who spent all afternoon at the hairdresser's, stout farmers with creased blue jeans and starched flannel shirts—pours onto the dance floor, the man takes the woman in his arms, and off they go, twirling around in wide circles as if caught up in a whirlwind. The horseless merry-go-round comes to life with a swirling blur of humanity.

Everyone has their own style. Some men have their elbows raised at an elegant 90-degree angle and guide their partners with an imperceptible featherlight touch; others grab on to their women like bears and practically throw them around the dance floor. Still others—laborers, perhaps, who slug through their daily motions with heavy tools and clumsy roughness—magically morph into Fred Astaires and sail around with astounding grace and agility. Some couples (such as Fabrizio and Marilena, who have recently taken dancing lessons) execute their moves with focused precision and seriousness, while others seem totally nonchalant, as if just out for a brisk walk. Some are old married couples who have obviously been doing this for a lifetime, their moves and cues as thoroughly predictable as the many other personal habits to which each has become accustomed over the years, while others are more awkward and tentative, emitting an almost palpable electric charge; perhaps they are not really a couple at all, perhaps they are dancing together—maybe even touching each other—for the very first time. There are children mimicking the adults, women dancing with women, and smoking teenagers ready to let loose when the band changes to rock and roll. Everybody has a good time; even the old people who can't dance anymore still love to watch the action from the sidelines, and the perimeter of the dance pavilion is now packed with plastic chairs. Gigi, Ivana's next-door thirty-something neighbor who has been confined to a wheelchair for much of his adult life, is there watching too.

I suspect there was a time when dancing at a summer festival was a rare and cherished opportunity for men and women to make contact, especially physical contact. Now things are different; every teenager has a car and *preservativi* (condoms) are sold in vending machines on the street outside most pharmacies. Dancing at festivals remains a solid fixture, but *ballo liscio* is teetering on the verge of becoming a lost art, an exclusively old persons' recreation. I suppose this is understandable: the music is old-fashioned; you need to know the steps—the fox-trot, waltz, tango, polka—and how to guide your partner around a crowded dance floor without stepping on toes or colliding with other dancers; and it is undeniably sexist: the man leads, the woman follows. But it looks fun; funner (and certainly more romantic) than the contact-free, free-for-all gyrating of the post-sixties. Perhaps *ballo liscio* will make a comeback. I resolve to take some lessons before the next *festa* season rolls around. In the meantime, however, I content myself with "modern" dancing to old disco music and joining in the hottest group dance of the season, something about a blind *capitano* with an infectiously upbeat Latin rhythm and a whole series of wacky moves I am quickly becoming familiar with. This, after all, is summer in the Langhe, *Ferragosto* at the Festa della Bussia, and the only real requirement is to have fun!

· RECIPE #18 ·

SUGO DI CARNE FESTA DELLA BUSSIA

Festa della Bussia's Famous Meat Sauce

EACH YEAR, A day or two before the *festa* is set to begin, three women—Wilma of the Bussia Sottana, Tiziana of the Bussia Soprana, and Chiara of Monforte *paese*—gather at what was once the old school to make meat sauce.

In a huge pot over a low flame they prepare a massive quantity of sauce with over one hundred pounds of meat; it must be enough to last all five days and sufficient to cover the mountains of ravioli that are served each night. And this dark, rich *sugo* steals the show; good as the meat-filled cushions of ravioli are, they provide little more than a soft pillow for the flavorful sauce.

"There's almost a case of Barolo in that sauce, you know," says Wilma in response to my query.

"Ah! Is that what makes it so dark?"

"Well, no, not really. What makes it so dark is that it cooked gently from twelve in the afternoon until well after midnight. You'd be dark too if you spent all that time bubbling away on the stove!"

"The Festa della Bussia is not famous for its ravioli," someone proclaimed at the most recent *festa*. "It is famous for its *sugo*!"

Here (scaled down considerably) is Wilma's recipe:

INGREDIENTS

½ cup olive oil

1 medium onion, peeled and
    finely diced

1 carrot, peeled and finely diced

3 ribs celery, finely diced

3 cloves garlic, finely chopped

1 pound ground pork

1 pound ground veal (or ½
    pound ground veal and ½
    pound ground beef)

Coarse salt to taste

1 handful each of fresh sage,
    fresh rosemary, fresh
    oregano tied in a bouquet

2 generous glasses (about 2
    cups) red wine (see note
    below)
3 cups tomato puree
2 tablespoons tomato paste
1 cup tomato juice

Pinch of crushed red pepper
    flakes and/or black pepper
    to taste (optional)
2 tablespoons fresh parsley,
    coarsely chopped

1. Heat the oil in a large heavy-gauge saucepan, add the chopped vegetables and garlic, and sweat on a low heat until they begin to turn soft and transparent (approximately 6 minutes).

2. Raise heat to medium, add the ground meat little by little, mashing with a wooden spoon to prevent any lumps from forming.

3. When the meat begins to turn opaque, raise the heat to high, add a healthy pinch of salt, the herb bouquet, and the wine, and continue cooking on high heat until the raw alcohol flavor burns off (approximately 10 minutes).

4. Stir in the tomato puree and the tomato paste. Lower the heat and allow the sauce to cook, stirring occasionally, for at least three hours. If it becomes too thick and dry, add tomato juice as necessary.

5. Add additional salt, if necessary, and pepper to taste. Stir in the fresh parsley and serve with meat-filled ravioli or your favorite pasta.

YIELD: MAKES ABOUT 2½ QUARTS OF SAUCE (LEFTOVERS MAY BE FROZEN FOR FUTURE USE).

———————

NOTE: For the *festa* sauce, they use Barolo wine because this is the Bussia and they have a lot of it around, but any decent red wine is fine.

WITH THE CONCLUSION of the Festa della Bussia, we have made it over the hump of the seemingly endless summer from hell. Fabrizio was right. It didn't rain on the festival, or on any other day for that matter; the heat, the intense sun, and the drought have continued unabated. We survive, however, by hanging out at one of the two nearby public pools and by taking frequent trips up into the mountains, where it is relatively cool and fresh.

Throughout the summer, I continued to help Fabrizio in the vineyard. But with time on my hands and ever-weakening dollars in my pocket, I seek out other work as well. Sometimes I help out at the restaurant in Castiglione, also called Le Torri, where Bruna works. But I much prefer the new experience of working outdoors among the vines.

At the Festa della Bussia, I had bumped into a local producer, actually a brother and sister team, who said they might be needing some help with the coming harvest.

Tiziana and Marco Parusso are the fourth generation of a farming family, but the first really to vinify and bottle wine under their own label; their father, Armando (whose name is still on the label and whose picture sits on a prominent shelf in the kitchen), made wine that was sold mostly *sfuso,* in bulk. The family home and cantina is located right in the Bussia, as are some of their best vineyards. They farm about twenty hectares of vines, half of which they own and the rest they lease on a long-term basis.

Several days after the end of the *festa,* I set out with their crew at 7 A.M. in the back of a tractor trailer. It is quiet, and though early, already quite warm. When we reach our destination, the waking vineyard is moist with dew and the sun rising up over Serralunga makes the ruby red grapes sparkle. The fat clusters protrude into the rows, making it difficult to pass. But not for long.

Dressed in short pants and T-shirts, carrying plastic bags of water and sandwiches wrapped in aluminum foil, we might resemble a harmless pack of picnickers. In reality, however, we are more like a group of raiders about to launch an early-morning surprise attack. We are here to do some damage.

When the tractor rolls to a stop under the shade of a tree, we pile down, fan out through the vineyard, and prepare to begin.

"You want me to do what?" I ask with barely concealed shock and surprise.

"Take your clippers, cut off these bunches, and throw them on the ground," says our leader, Enrico, with just a hint of impatience. "Like this."

He attacks the first bulging vine, clipping wildly and fiercely but also with ruthless precision, letting the severed fruit fall heavily onto the earth around the vine. In just a few moments the green grass is littered with plump red grapes; it seems that there is actually more on the ground than left on the vine.

"See?" he says, victoriously munching on a bunch of grapes. "It's easy. And it's fun! Now you try."

"Uh, okay," I respond, a little less enthusiastically. "Just cut 'em off and throw 'em down, right?"

"Yeah, that's right—well, basically. Look at each branch. We want to leave one bunch on each branch, the best bunch, and preferably the one closest to the bottom (usually that's the best one anyway because it's closest to the base where the sap comes from). See this branch here? There's three bunches on it. They'll never all ripen—not the way we want them to anyway—and certainly not in a shitty year like this one. So we cut off these other two—*sak! sak!*" He slowly positions his clippers and snips off the upper two bunches to demonstrate his point. The one that's left will be better. Less fruit, better quality, you understand? If there's a vine with fewer branches, like this one with only five, you can leave two nice bunches on one of the branches, but no more than eight per vine, okay?"

"Yes," I said. "Makes sense. But it's a shame about all those lovely grapes lying there on the ground like that. Isn't there anything you can do with them?"

"Well, *you* can come back later and collect them and make grape juice if you want to." He and some of the other workers get a good chuckle out of this. "But they are no good for making wine, not good wine anyway. They're not mature, and there's not enough sugar to ferment. And wine—*good* wine—is what we are here to make, right? So how about we stop talking now and get to work."

This last sentence was pronounced with a note of humorous camaraderie, but there was no question mark at the end. It was time to get to work.

Though this was my first time in Parusso's vineyard, I have spent enough time with Fabrizio in Le Munie to develop an almost protective, paternal feeling toward this fruit. For months the vines have been groomed and preened, the budding shoots coddled, the branches untangled and supported, the clusters nurtured. In the sun, snow, rain, and shadow of storm clouds, early in the morning till after dusk, we have toiled, sweated, strained muscles, and in some cases, spilled blood. And now that the grapes have turned bright red, grown big and fat, and are close to reaching maturity, we do the unthinkable. Like brutes on a rampage, we move up and down the rows raping and pillaging, murdering the vines' defenseless offspring in their near-prime.

It's called *diradamento* (reducing the yield) or *vendemmia verde* (green harvest), a placid name for what is on the surface at least a thoroughly violent act. (In Piemontese, they refer to it more succinctly as *buta giù*, or "throw down.")

There's really no contest; in this case, the plants don't even get in a slap. Vine after vine, hour after hour, day after day, we cut off all the clusters but one on each branch and throw them down on the ground. Occasionally, I take a bite out of the bunch before tossing it nonchalantly away, the sweet/sour juice running down my face, my hands red and sticky as if dripping with blood. I am satiated but keep taking bites, nonetheless; take a mouthful, drop the rest, suck out the juice, and spit out the bitter seeds and tough skin. In the end, the once-pristine grassy carpets are thickly littered with fallen fruit.

Reducing the yield during the final growth spurt—that is, limiting the number of grape clusters on each vine by removing a significant amount of fruit about a month before harvest—focuses all the energy and resources into the remaining bunches, thereby significantly intensifying their flavor and concentration. It mostly comes down to a King Solomon–like choice of quality over quantity.

This, of course, can sometimes be a difficult choice to make, especially for a thrifty old farmer who toils long and hard to make his plants as productive as possible. To some extent, the priorities of the grape grower (who doesn't make wine) and the winemaker (who doesn't grow grapes) are at odds; the first wants to get as much fruit as possible from his land in order to have more to sell, whereas the latter wants the best-quality fruit (at the lowest possible price) in order to make the best possible wine. The discrepancy is somewhat mitigated by the fact that better-quality grapes are *worth* more since they make a better-quality wine (which, in theory at least, commands a higher price). The end result is that in many cases a farmer can actually make more money by producing fewer grapes.

When, as in the case of Parusso, the grape farmer is also the winemaker, the problem takes care of itself. Parusso strives for a reasonable yield of top-quality fruit, which for them generally means one bunch per branch, or about one kilo and a half, of grapes per vine. That translates roughly into one bottle of wine per vine, even less perhaps for a wine like Barolo, which ages for two years in barrels before being bottled; during this period, each barrel loses about 8 percent of its volume to evaporation, a loss euphemistically referred to as the "Angels' Share."

Often, when the farmer is not the winemaker (like Fabrizio, at least for the time being) the person to whom he will sell his grapes will probably have a major say in how much is ultimately harvested, whether *diradamento* is done and if so how much. In Barolo, maximum yields are strictly regulated by law. You can harvest a maximum of 80 *quintali* (8,000 kilograms) per hectare of nebbiolo grapes from within the zone to make Barolo; nearly everyone, however, obtains less than the maximum (50 to 70 *quintali* is average for a quality producer).[1]

Midway through the growing season, but well before the harvest, the *consorzio* sets a standard benchmark price for grapes for that year.

On that basis, individual negotiations will take place between grower and buyer; according to the quality of fruit and desirability of a given vineyard location, a price at, above, or below the set price will be agreed upon. The buyer, in effect, agrees to purchase the production of a specific area of vines based on the maximum yield allowable by law. He can then determine the actual yield he wants (as well as other issues of vineyard management like chemical treatments). If he wishes to reduce the yield, he will simply end up paying more per kilo for higher-quality fruit.

In most cases, these arrangements between farmer and winemaker are long-standing relationships that go on year after year—sometimes generation after generation—even though many of them are based solely on a handshake. Official, written contracts are rare.

Toward the end of August, I learn that the grapes hanging heavy from the vines of Le Munie—the grapes that I have helped tend at every step since the first pruning way back in January and that are now getting ever closer to reaching maturity—have already been sold; they will go to Franco Conterno, who has a cantina just down the road in the Bussia Sottana. He has vineyards nearby; he will help with the harvest and use Fabrizio's grapes in his own Le Munie Barolo cru.

While I continue throwing down grapes with the Parusso crew, I go back to help Fabrizio as well, but we don't need to do much *diradamento* in Le Munie. The quality of the grapes is fine but the quantity is already quite low because the vines are old. That can sometimes be good, since mature vines generally produce better (if fewer) grapes. But because these vines haven't been maintained meticulously over the past twenty-five years since Francesco's death, they are tired. Fabrizio has decided that this first harvest will be his last—at least for four years or so. After these grapes are picked, the old vines will be ripped out, and the entire vineyard completely replanted.

· RECIPE #19 ·

RISOTTO AL BAROLO

RISOTTO IS MADE throughout northern Italy, but it is most at home in Piemonte and neighboring Lombardia, where rice is grown in extensive paddies on both sides of the Po river due west of Milan.

There are three main varieties: arborio, cannaroli, and vialone nano. While each is slightly different (cannaroli, for example, is usually the hardest and most resilient), they are all plump, short-grained, and starchy. You can make many tasty rice dishes with other types of rice, but you can only make *risotto* with one of these Italian varieties.

Risotto is extremely versatile, but the process of making it is almost always the same: the rice is first "toasted" (lightly sautéed) in butter to create a shell around each individual grain. Hot stock is then added little by little, stirring often (some insist constantly), until the rice is cooked. As the rice cooks, it gives off its starch, making the whole thing appear creamy. It is often finished with a dollop of butter and grated Parmigiano-Reggiano cheese.

Risotto takes anywhere from 20 to 30 minutes, depending on which type of rice is being used and how much is being made. Steer clear of any shortcuts or instant versions: real risotto is worth the wait, you can nibble an antipasto while you stir.

INGREDIENTS

2 tablespoons butter
1 quart chicken stock
1 cup Italian arborio, cannaroli,
    or vialone nano rice
1 onion, peeled and diced
1 cup plus 1 tablespoon Barolo
    wine
Generous pinch of salt

1 tablespoon grated
    Parmigiano-Reggiano
    cheese
Fresh black pepper to taste
1 teaspoon coarsely chopped
    fresh Italian parsley
¼ cup red seedless grapes,
    halved (optional)

1. Melt half the butter in a heavy-gauge sauce pot. In another pot, bring the chicken stock to a boil, then lower the heat to a simmer.

2. Add the rice to the butter, and, stirring with a wooden spoon to coat each grain, "toast" until the rice begins to sizzle (do not, however, allow it to take on any color whatsoever).

3. Stir in the diced onion, and cook until it begins to soften, about 2 minutes.

4. Add the cup of wine, and allow most of alcohol to burn off, about 2 minutes. When the rice has absorbed the color of the Barolo and the wine is almost cooked away, lower the heat to a simmer.

5. Begin adding the hot chicken stock, little by little, to just barely cover the rice. When the stock has cooked down and is almost completely gone, add more stock. Add a generous pinch of salt.

6. Continue in this way until the rice is almost cooked, about 18 to 20 minutes. (Toward the end of the cooking time, add the stock very judiciously. If too much stock is added, the rice may become mushy, water-logged, and overcooked.)

7. Test the rice for doneness: the rice should be creamy and homogeneous, yet the grains should be individually distinguishable, puffed up but not burst open. Take a grain and bite it in half: inside there should be just a tiny speck of whiteness at the center. This means the rice is perfectly cooked *al dente*—chewy and resilient, but not dry and crunchy.

8. Add the tablespoon of wine. Turn off the heat, stir in the remaining butter, the Parmigiano-Reggiano cheese, black pepper, and chopped parsley. Add more salt if necessary.

9. Divide the risotto into four bowls, sprinkle the grapes over the top, if using, and serve.

YIELD: SERVES 4 AS A *PRIMO PIATTO*.

———

NOTE: You may wish to top the rice with a piece of Gorgonzola cheese for each person to stir into their rice. You could also serve this risotto with a ladle of the Bocconcini (page 28) or Fonduta (page 102) in the center or as an accompaniment to some other main course.

25 · TURNING OF THE TIDE

THE THIRD WEEK of August I do my first real harvest with Cantina Parusso. We are out to pick sauvignon bianco. Parusso is one of a small number of producers who grow white grapes in the Barolo zone, and one of a very few to grow sauvignon (the more common white varieties are the chardonnay, the indigenous favorita, the sweet moscato, or the rare arneis).

It is extremely early; no one can remember ever picking in August before. But the grapes are ready: they are plump and juicy, the sugar level is unusually high, and the pH balance correct. You can taste the *rightness* in the grapes themselves (or at least experienced farmers can), but the sought-after chemical attributes are nevertheless confirmed by laboratory analysis.

This mini-harvest of white grapes lasts but a few days and is little more than a warm-up drill for the bigger one that will follow. Though it is still hot as hell, the growing season is dragging slowly but surely to a close. Soon the pace will quicken. The endless parade of *feste* and the lazy long hot days of patiently watching the vines grow are about to explode in the frenetic activity of *vendemmia*. The quality and quantity of the harvest, however, remains to be seen, and the air is thick with anticipation.

Notwithstanding all the advancements of technology, the sophisticated meteorological forecasts, and the modern techniques of vineyard management, in the end the quality of the harvest still depends on the same age-old factors: the weather, the ineluctable forces of nature. This is an important lesson to learn, especially for someone like me who comes from a place like New York—in many ways, the antithesis of a natural environment. In New York, a big rainstorm mostly means it will be hard to find a taxi.

Here in Barolo it is different.

As every winemaker worth his salt knows and freely admits, wine is made first and foremost in the vineyard; the goal in the cantina is mostly not to mess it up. By the same token, if the fruit itself is not great, little can be done to make great wine out of it. Man is a collaborator in this enterprise—that is what I've been doing for the past eight months, *collaborating;* pruning, supporting, weeding, lending a hand, helping the vines absorb and transform the energy of their environment—but only a collaborator. Nature remains the big boss, and a sometimes difficult, unpredictable one at that.

I could not have picked a better first season if I'd tried: 2003 has been a weird year, an anomaly, and no one can remember a season quite like it.

It has been brutally hot and dry all summer long, and the vines have been stressed. Many of the young plants (those less than five years old) have been wiped out, not having had a chance to develop the extensive root systems necessary to find water deep under ground. And many of the grapes on the mature vines have been literally toasted by the intense sun; those with some shade—bunches in the back or those on vines whose farmers did, for whatever reason, less leaf clearing—fare better. Quantity will be down, both because there are fewer grapes and also because the fruit itself is drier than usual, less juicy.

But they're holding their own. Wine grapes, it turns out, actually *like* a certain amount of stress. It builds character. It is much better for them to have too little rain than too much (as in 2002), and the intense sun is actually beneficial for early-ripening grapes like dolcetto and barbera that sometimes have a hard time reaching the desired sugar levels. Due to the extreme dryness, fewer chemical treatments are required, and the clusters are turning dark and plump with barely a trace of rot or mildew.

The tide of opinion is starting to change; perhaps it will turn out to be an okay season after all, though no one will come out and say it. The Piemontese are stoic, hardened, superstitious farmers, and they don't want to jinx it. Plus, they know all too well that anything can still happen. "In any event," said one, "there's nothing you can do about it. Nature does what she wants; whatever happens will happen. In a couple of weeks we'll know for sure."

*　　*　　*

All summer long—ever since I first moved here in May, in fact—I'd been looking for an apartment to rent. It had not been easy. In rural Piemonte there just aren't many apartments to begin with. And owners of the few rentable places that do exist generally prefer to rent them out to tourists on a short-term basis. There are, I discovered, many empty houses around, but, unless the owners are strapped for cash—and most of them are not—they'd rather let them sit empty than take on the risks and responsibilities of being a landlord. (In Italy it is next to impossible to evict tenants, even if they don't pay the rent or cause some other sort of trouble.)

It would surely have been easier to find a place in Alba where apartments are more plentiful. But that's not where I wanted to be: Alba is a city—a small city, perhaps, but a city all the same—and I didn't come here to live in a city. My heart was up in the rolling hills of the Langhe. It might only be a ten-minute drive on the *superstrada,* but it seemed a world away. And besides, I didn't have a car.

I asked everyone, scoured local newspapers, and even found a few apartments to look at, but none of them were right. I had an image of the place I wanted; a charming old stone house with timbered ceiling and fireplace, great light, lovely views, and (hey, why not?) a little garden. Plus, if at all possible, I wanted to be in Castiglione; I knew everyone in town by now, and it was starting to feel like *my* place. So I sat tight in the hotel while the summer dragged on and I continued my quest.

Finally, just as the ripening grapes were about ready to burst, my persistence paid off. The young couple that runs the restaurant Le Torri had decided to buy a condo in La Morra, leaving their rare rental apartment available—right in Castiglione! Wasting no time, I arranged to take a look at it and Ivana came too.

The apartment is located in a renovated yellow house at the foot of town, occupying a full floor above La Coccinella (The Ladybug), the *other* bar where workers go for morning coffee. It consists of a decent-sized bedroom and large *soggiorno* kitchen (most kitchen spaces here are not only eat-in but live-in, with extra room for a couch and people to hang out), a smaller second bedroom or study, a full bathroom, and a separate room for laundry. The apartment has two balconies in front, a huge *terrazza* in back, and even a sizable overgrown area for a garden, should someone ever choose to take on the project. It gets great light with the morning sun pounding the terrace, and the northwest-facing balconies look out across the Alba-Monforte road over rolling vineyards to La Morra and the Alps beyond.

At first, the idea of being so close to the road and the café sounds wafting up through the floor bugged me a little. *"Ma sei pazzo?"* Ivana said. "How can somebody from New York possibly complain about a few passing cars?" (When I was in New York and we spoke on the phone, the constant roar of traffic, the car horns and sirens, the ubiquitous sounds of the city, inevitably drifted up eighteen floors and into our conversation.) "And the people downstairs don't bother me at all; I *like* the sounds of life and activity." In fact, Ivana unequivocally approved of the apartment; aside from all its other amenities, it is newly renovated and eminently cleanable.

Though renting an apartment here had always been part of the plan, it seems that, now it is time to actually do so, I am resisting, as if I'm *trying* to find reasons why the place is not acceptable. But, while it may not be perfect, there is no denying that this is the best possibility to come along. In any case, I really need a place: I am here in Italy and my home in New York has been rented out. Moreover, the seemingly endless summer is, in fact, rapidly coming to an end, and I will soon have to be out of the hotel before the autumn tourist crunch. So I decided to take it.

From that point, things happen quickly and easily. I sign a lease and have the utility accounts transferred into my name. The apartment is unfurnished—as almost all apartments are, here as well as in New York—so Ivana and I make several forays to the nearest Ikea in Torino, where we select a bed (*matrimoniale*, of course), linen, a lamp or two, and a huge four-season armoire. The size of this armoire scares me—it occupies almost an entire wall—but there are no closets in the apartment (in Italy built-in closets are almost nonexistent). And besides, with extra storage space, should anyone else ever decide to leave some stuff, there will be room.

In Italy, I learn, "unfurnished" also means no kitchen. Evidently, kitchens here are considered personal property, and people lug them

around from place to place when they move, disconnecting and connecting and modifying them as necessary. There had never been a
kitchen in this apartment (the previous occupants ate all their meals in
their restaurant), but the necessary gas, electric, and plumbing outlets
were there, ready and waiting, popping out of the tiled wall. And pur-
chasing *una cucina* is fairly easy; you can do it piecemeal if you want,
but typically they are all-inclusive combos of fridge, sink, oven, coun-
ters, cabinets, and drawers. You just measure carefully, making sure the
position of each component lines up correctly with the outlet, and
make your selection. I had never purchased a kitchen before; it necessitated difficult decision-making—color, style, finish, disposition of cabinets, dishwasher or no—and several extra trips to Torino. But eventually,
I just picked one. With that, the task was accomplished and my little
home in the Langhe on the verge of becoming a reality.

By this point, the summer was almost over and, except for periodic
visits to one of the nearby swimming pools, we hadn't been near water. Now, I am happy to be here: I love the hills and vineyards and
forests of my new neighborhood, and our trips to the mountains have
been a welcome respite from the heat. I have already spent much time
outdoors, and my skin has turned quite dark—at least from the waist
up, the characteristic vineyard worker's half tan—from hour upon hour
among the vines. But over the years, summer has come to mean Cape
Cod, and I long for the soothing expanse of the sea and the smell of
salt. I have, basically, just gotten here; I am just beginning to settle in
and harvest is just around the corner. Going back to the States and out
to the Cape is out of the question, and, anyway, I don't want to leave
Ivana. So we do the next best thing: we get in the car and drive south to
the sea.

There is something marvelously euphoric about heading to the sea,
especially in the summertime; an almost instinctual, gravitational pull,
like migratory birds on their natural seasonal rounds. Plus, I have never
traveled with Ivana. So as we throw the basic necessities into a shared
suitcase—sunscreen, swimsuits, shorts, and a light jacket for cool
evenings—and set out on the last Friday of August just as dawn is
breaking, I feel a giddy sense of lightness and anticipation.

Leaving Piemonte, one of the Italian peninsula's few landlocked regions, for neighboring coastal Liguria, I watch the gradual transformation of the landscape unfold along with the rising sun; vines yield to olive orchards, populated hills change to cliff-punctuated pine forests and then, finally, palm trees at an ever-decreasing altitude. We reach water in time for morning cappuccino and then head south along the curvaceous Riviera dei Fiori road to the beautiful old hill-town of Cervo rising up vertically from the sea.

It was an excellent choice (brava, Ivana!): Cervo is one of the less-touristy places along the Italian Riviera and one of the few not to smack completely of Summer Holiday. As we check into our rather dumpy but perfectly sufficient hotel at the foot of town, I am surprised at the preponderance of young Asians lugging instrument cases around and, on the way upstairs, even more surprised at the sounds of music spilling out of rooms into the hallway. It reminds me of my long-lost conservatory days and gives me a pang of longing for the piano along with a slight sense of guilt by omission, as if I should be practicing too. (As we later learned, there was an international music workshop going on, and the kids were preparing for their big end-of-season concert.) We dump our stuff and head right out, strolling hand in hand through the narrow twisting streets of the old village perched high above the sea. Echoes of the practice-room cacophony stick in my head, however, blending with the real sounds of leather sandals on smooth stone and caged songbirds in windows. (I resolve to get myself a piano when we get back home.)

Liguria is not Cape Cod and doesn't have the vast, sandy stretches of the National Seashore. Beaches are few and far between—rocky promontories or wooden decks sticking out above the water are the norm—and most of them are constantly raked and clogged with rows of reclining chairs and food concessions. You pay for admission and your chair. On the rare public beaches, people pack themselves in like slippery sardines and lie practically elbow to elbow on the well-trodden, cigarette-butt-filled sand while Moroccans tiptoe over them selling trinkets and chilled coconuts.

But it doesn't matter. Each day we find a spot by the water, plop down on our rented chairs, and there, as close to naked as decency permits, sizzle in the sun. Ivana has a very strict regimen: on goes the blue bandana to keep her hair off her face, and off comes the bikini top; fifteen minutes on one side, flip over, fifteen minutes on the other; turn the chair as necessary to keep in direct head-on alignment with the sun and No Touching Allowed! other than fingertips—it ruins the balanced tan.

This routine is broken only for lunch and the no-touching rule mercifully applies only during sunning. When late afternoon rolls around, we gather up our things, head back to the hotel, and, in our dingy-but-perfectly-sufficient room, press our throbbing, well-oiled bodies together in a deep, sea-scented act of love with the plaintive melody of a Bach solo cello suite softly serenading us through the wall. We then shower, dress, and go out for dinner.

All that sun (and sex) works up an appetite. We sit on outdoor patios dripping with flowers and greedily consume vast amounts of seafood and pesto washed down with cool, crisp Ligurian wine, while the fragrant breeze blowing off the Mediterranean makes our warm flesh tingle and distant lights along the coast sparkle like fireflies. Then we drive to a nearby town such as Alassio or San Remo to eat ice cream and walk up and down the brightly lit streets packed with throngs of other vacationers, who, like us, seem swept away in an end-of-season frenzy.

In an instant, it seems, our weekend—and the summer—is over, and it is time to go. I leave with a touch of moping sadness, like a kid going back to school. But I am also full of anticipation; I am excited at the prospect of returning to the vineyards for the impending harvest and of settling into my new home.

IN A SENSE the next two recipes don't belong in this book. They are unequivocally not Piemontese. In fact they represent a whole other world, another reality. They sing of the summer, the sea, and the perfumed, flower-filled Ligurian coast, site of our beach escape. But I include them for several reasons. They're great simple dishes to prepare and may provide a refreshing counterpoint to the rest of the robust, meaty, strictly regional sub-Alpine dishes included in this book. I also think they illustrate a worthwhile point: sometimes in order to understand or appreciate the regionality of a given region, you need to leave it.

· RECIPE #20 ·

SPAGHETTI ALLA PESCATORE

Spaghetti, Fisherman's Style

THE ACTUAL CONTENTS of Fisherman's Spaghetti will vary considerably based on what the fisherman happened to catch (either at sea or at the market) and where the fishing took place: the Mediterranean, the Atlantic, or the Pacific will yield different though equally valuable treasures. The important thing is to use a variety of seafood and have it be as fresh as possible, as if a mix of sea creatures had just been lifted from the sea and dumped into a big frying pan.

Most of the cooked shellfish remains in the shell. This makes for a dramatic presentation as well as something of a dining challenge: you have to pick the clamshell up to scoop out the meat, pull the head off the shrimp and suck on it to extract flavor, gnaw on crab legs or claws. It is messy work best done in shorts or bathing suits, but fun and worth the trouble. (It's also much better than having had someone else's hands all over your seafood, intervening between you and your catch.) Not to fear: the briny mess will all come off in a warm shower or an impromptu dip in the cool sea.

INGREDIENTS

1 pound spaghetti or other dried
    pasta
Coarse sea salt to taste
¼ cup extra-virgin olive oil
A combination of fish and
    shellfish such as:
½ pound calamari (or fresh
    cuttlefish), cleaned, rinsed,
    and cut into rings and
    tentacles

¾ pound small clams (called
    vongole veraci in Italian),
    such as littlenecks or
    cockles, washed
½ pound mussels, washed
½ pound scallops
8 scampi (Mediterranean
    crustaceans) or ½ pound
    fresh shrimp
2 small blue crabs (or substitute
    ½ pound crab meat)

1 pound fresh fish, such as
   salmon, tuna, monkfish,
   and/or halibut, cut into 1-
   inch cubes
1 clove garlic, quartered and
   thinly sliced
½ cup white cooking wine
1½ cups fresh peeled, seeded,
   and diced tomato

½ cup tomato puree or 1
   teaspoon butter
½ teaspoon hot pepper flakes
   (optional)
1 bunch of fresh Italian parsley,
   leaves picked, washed,
   patted dry, and coarsely
   chopped

1. Make sure all of the seafood and the other ingredients are cleaned, cut, and ready to go. As long as that is the case, preparation of this dish shouldn't take much more time than that required to boil water and cook the pasta.

2. Place a large pot of water on the burner and bring to a boil. Season the water with salt to taste, add the pasta, and cook it.

3. While the pasta is cooking, heat the olive oil in a large skillet. On a high heat, add the calamari, the clams, the mussels, the scallops, the shellfish, and any other seafood that takes more time to cook. Sauté for 2 to 3 minutes.

4. Add the cubed fish and the garlic and give the contents of the pan a toss (you can use a wooden spoon if necessary, but be careful not to break the fish!).

5. Still on high heat, add the white wine to the pan and let the alcohol burn off (about 3 minutes). Add the diced tomato, cover, and lower heat to medium.

6. When the clams and mussels have opened (usually about 4 to 5 minutes), remove the cover. Add the tomato puree or butter to help bind the sauce. Add the pepper flakes (optional), additional salt if necessary, and finally the chopped parsley.

7. When the pasta is cooked *al dente* (approximately 12 minutes), strain and add to the seafood, tossing to combine thoroughly. Pour the contents onto a large platter and serve. (Be sure to provide an empty dish for shells!)

YIELD: SERVES 6 TO 8 AS AN APPETIZER OR 4 AS A MAIN COURSE.

· RECIPE #21 ·

PESCE AL SALE

Whole Fish Baked in Sea Salt

FOR PEOPLE WHO love seafood, it doesn't get much better than a whole, flappingly fresh fish baked under a crust of coarse sea salt. The utter simplicity of this cooking method takes nothing whatsoever away from the fish itself. (It won't add anything either, however, so if you can't score a really fresh one, just throw it on the grill instead).

Though you might expect the end product to be overly salty, it is not: the salt solidifies as it cooks, forming a protective seal that keeps all the moisture and flavor in the fish itself; because the salt doesn't dissolve, it doesn't overly season the fish.

Besides a moist, flavorful fish, this preparation makes for a marvelous presentation: the whole fish is brought from oven to table. As the crust is removed, like a veil lifted from some marine masterpiece, the fragrant steam teases your nostrils even before the succulent sea creature's flesh melts in your mouth. A squeeze of lemon and a drizzle of delicate Ligurian olive oil is all it needs.

INGREDIENTS

1 fresh whole fish (such as red snapper or sea bass), approximately 2½ to 3 pounds, gutted, scaled, and cleaned (gills removed)
1 lemon cut into wedges
1 small bunch of fresh herbs (rosemary, thyme, bay leaf)
1 pound coarse sea salt

Optional: ¼ cup mixed spices such as peppercorns, cardamom, cloves, star anise, allspice (see note below)
1 large rectangle of heavy-duty aluminum foil

1. Preheat oven to 400 degrees. Rinse the fish under cold running water and pat dry. Place the fresh herbs and one lemon wedge inside

the cavity of the fish. Fold the foil in half to create a smaller rectangle that is just slightly larger than the fish itself.

2. Place the fish on the foil with the head to the left side (this will make it easier for a right-handed person to filet); there should be about a 2-inch circumference of foil around the fish. Bring together the upper and lower corners of foil at the head and tail of the fish and fold over to secure. Fold the foil edges up around the sides of the fish and fold down if necessary. In the end, the foil should wrap snugly around the fish with a lip approximately ¼ inch all around.

3. Place the coarse salt and spices (if using) in a mixing bowl. Add a splash of water (no more than a tablespoon) to moisten slightly, mix well, then spread over the fish to cover completely, creating an even layer about ¼ inch thick.

4. Gently lift the fish-in-foil package and place it on a baking sheet. Put the fish in the oven (or onto a grill rack) and bake for 40 minutes. Remove from the oven and let stand in a warm place for an additional 5 minutes. (Note: Because the salt acts as a protective seal, it is difficult to overcook and dry out the fish.)

5. Place on a serving platter and bring to the table. Loosen the foil from the sides and lift the hardened salt crust off in one piece. Often the skin will come off with the salt; if not, gently peel it off, and, using a spatula, lift off the top fillet onto a plate. Remove the head and bones and then lift the bottom fillet onto another plate. Garnish with the remaining lemon wedges and serve with a drizzle of olive oil.

YIELD: SERVES 2 AS A MAIN COURSE.

NOTES: It is preferable to remove the gills as they can sometimes create a slight off-putting odor when cooked. Since removing gills can be difficult without a heavy set of cooking shears, it is best to ask your fishmonger to do it for you.

The spices are not absolutely necessary and don't really affect the taste of the fish, but I like to mix them into the salt for the aromatic effect they give prior to removing the crust.

PART IV

AUTUMN
BOUNTY

WHEN WE GOT back to the Langhe that evening, everything had changed. There was a long-lost lightness in the air and the stifling humidity was gone. The next day it even rained, not just another impotent misting, but a real (hail-free) downpour that thoroughly refreshed the parched vines. Overnight, it seemed, the prognosis for the vintage had done a 180-degree turnaround. The stone-faced farmers were ecstatic; a few of them even almost cracked a smile. "I have never seen dolcetto like this," one said. "And the barbera—*bellissimo!*"

In autumn the Langhe, always beautiful, comes into its prime. The landscape is finely etched; the Alps, in between the haze of summer and fog of winter, stand out crisply on the horizon, and the vineyards sparkle in the sharp clear light. The dolcetto vines, always the first to turn color, stand out distinctly from the rest, like brilliant red islands in a sea of green. Though it is still quite warm (the calendar still says summer), the quality of the air is different and carries a hint of the cooler days to come. After the long summer lull, the pace begins to quicken; farmers rev up their tractors and line up their picking teams.

The day after we got back from Liguria, I borrowed an old sky-blue, black-cloud-diesel-gas-spewing van with a cantankerous clutch, and we chugged to Torino to pick up the furniture. Then I checked out of the hotel once and for all, transferred my few possessions into the apartment, and assembled the bed, while Ivana put the finishing touches on her cleaning campaign. That night we inaugurated both the new bed and the new apartment. And the next day I headed back out to the vineyards.

The dolcetto harvest began that first week of September with the barbera right on its tail. It was incredibly early—usually the harvest doesn't start until late September or early October—but the grapes were practically screaming, "We're ready! Come pick us!" And we did.

This was my first real harvest. I had a preconceived notion of what it would be like—a bunch of backpack-toting, suntanned students from around the world happily tossing grapes into wicker baskets while singing old Beatles' songs, then sitting down to lunch at a long communal table under the shade of a vine-covered arbor—but it turned out to be quite different.

Here in the Langhe, most of the pickers are either full-time Macedonian workers or older retired Piemontese. The old people are locals who return each season to the same *azienda* to help with the harvest, like salmon returning to the same spot each year to spawn. At Parusso there is Nicola from Bra and Nicoletta from Roddino (both of whom also help out during the year with the *tòrse* and *scarzolé*); there is Beppe, a retired employee from the Ferrero chocolate factory in Alba, and Gigi, an old raspy-voiced in-law from Monforte. They have all been harvesting with Parusso for over a decade. At Azienda Cavallotto in Castiglione, where I also help pick, there is Domenico, whom I had often seen hanging out in Renza's or driving his little tractor back and forth to his own small vineyard, and Angelina, a retired employee from the Alba-based textile company Miroglio, who, when she's not working in the vineyards, acts as the town seamstress.

These old folks—*anziani,* as they are known—pick grapes for two reasons, to make some extra cash to supplement their fixed pension income and *because that's what they do*. It's an autumn ritual: it gets them out of the house or café and into the vineyards socializing with their neighbors and working again, if only for a few weeks. The work can be challenging physically—standing in the sun for hours, bending and straightening, walking up and down the steep vineyard paths hauling heavy crates of grapes—but these *anziani* seem to do just fine. Most of them have been doing it all their lives.

Occasionally people do break into song while picking, especially in the late afternoons, but they are mostly solo renditions of older folk or pop songs. More common is the constant back and forth banter of Piemontese—the *anziani* telling jokes and stories, exchanging news, commenting on the grapes—which forms a kind of background music for the work.

There are usually six to eight people on a picking crew, broken up into teams of two. As the only American, I am something of a novelty and usually end up fielding questions, telling my partner-for-the-day about New York City, giving my eyewitness account of 9/11, or trying to explain how I happened to wind up here. Each team works back and forth along the rows, up and down the vineyard. We use red plastic two-handled containers that resemble big rectangular milk crates, and pointy metal shears.

Starting at the head of a row, you place the crate between you, clip off all grape bunches within reach and then, with a mutual nod of assent, lift the crate from each side and move on down the row. When the crate is full, you tuck it up close to the vines (or, if the row is too narrow for a tractor to pass, carry it down to the aisle at the end) and start on a new crate. Angelina always makes sure to place the most perfect, most beautiful bunches on top to "make the *padrone* happy when he sees them." Each crate holds about 20 kilos (approximately 40 pounds) of grapes, depending on how full it is. Periodically, a tractor passes through the vineyard to pick up the crates, which are then stacked onto a larger tractor trailer and hauled back to the winery.

We start at seven in the morning when it is still comfortably cool and work until seven in the evening with an hour-and-a-half or two-hour break for lunch.

With no permanent job and no long-standing allegiance to any particular *azienda,* I am something of a grape mercenary, picking for whoever needs a hand. I work with a number of different growers, but keep gravitating back to Parusso. With twenty hectares to harvest, they offer the possibility of steady work; I like their tight organization and meticulous standards of quality, plus there is the appealing Bussia connection. On top of that, they invite me to lunch each day in their home/cantina, and their mother, Signora Rita, is an excellent cook.

Not everybody gets lunch in the house (those that live nearby go home while others eat a bag lunch outdoors in the shade). Inside is the core crew: Tiziana and Marco (when he's not on the road trying to sell their wine); the two young guys who are the *azienda's* key employees, Francesco (who has an enology degree), in the cantina, and Enrico (he went to agriculture school), who focuses on the vineyard work; Giulia, Tiziana's rambunctious prepubescent daughter; Stefan, a young, gentle Aryan giant attending enological school in Austria and here for a summer *stage* (internship); and Giovanni, the big guy I noticed every night at the Festa della Bussia throwing his partner (Signora Rita) around the dance floor, but whose name I didn't know until now.

Giovanni is from Mondoví. He doesn't appear especially muscular, but he is incredibly strong, has a rich baritone voice, and deep wrinkles in his big-featured face. Mostly he drives the big tractors and heavy machinery, but sometimes he'll jump down to help pick too. Giovanni works like he dances—with big sweeping motions, energetic verve, and forward momentum; the few times I got paired picking with him, I had a hard time keeping up as he bent down over the vines and raced down the rows, pulling off the grapes and tossing them in the crate like a bear pawing the iron bars of his cage.

And he drinks the way he works. Giovanni likes wine; he prefers dolcetto to Barolo, but will drink what's there, and at Cantina Parusso, of course, there's always something there. He'll put away upward of a bottle of wine at lunch, then get up from the table with a grunt of contentment and return to work, maneuvering the huge tractor trailer around the precariously steep and narrow dirt roads with agility, precision, and grace. (Giovanni, I soon learn, is not just another employee; he is Signora Rita's companion, almost a member of the family. They met about six years ago at the Festa della Bussia.)

With Parusso, I get in on the second half of the dolcetto harvest. After that, around the middle of September, we go pick barbera in their vineyards in the Ornati section of Monforte for about a week. And then there's a lull.

The harvested grapes are lovely indeed—the sugar level is nearly off the charts—and people's excitement about the powerful, superior wines they will make is well founded. Everyone is happy and relieved; at least they will have *something* this year. But here in the Langhe, nebbiolo is the most important grape, Barolo the undisputed king. It's partly about money: these grapes and the wines made from them carry the highest price tag. But it's not just that. Nebbiolo is like a favorite son; every Piemontese "father" wants him to do well, make a good name for himself, put his best foot forward. The international reputation of the region hangs on its Barolo and Barbaresco, even if what people here mostly drink is dolcetto.

According to the chemists' data, by mid-September the nebbiolo is ready to be picked, despite the fact that it is a month ahead of schedule; the grapes have reached the required level of sugar (measured on the Brix scale) and the pH balance is correct. But because of the extreme heat, they have ripened fast, some would say *too* fast. The grapes may be ripe, but they are not really mature; they have attained the correct chemical properties but not the balance, depth, and complexity necessary to fully express the *terroir* and make really good wines.

Pick or wait? It's an old farming gamble, a nerve-racking conundrum. A good portion of the year's income is hanging there on the vine and a late summer/early autumn storm could still come and destroy the entire nebbiolo crop. Some people don't want to risk it, especially after what happened last year, and start picking. But most prefer to hold out as long as they possibly can. Nebbiolo loves the warm balmy days and the now quite cool evenings that were absent all summer long. It also loves its namesake fog (*nebbia*), which has recently reappeared in the late afternoons.

The grapes might not be getting any *riper,* but the longer they remain on the vine, soaking up everything around them, they're getting *better*. We wait.

Finally, near the end of the month, Marco makes the call:

"Okay, these grapes have done all that they're going to. It'd be pointless to let them hang any longer. Let's get 'em in here!"

We dive into the nebbiolo vines in the Mariondino vineyard in Castiglione, a three-hectare parcel that takes a full two days to pick. Marco sets the overall tone of the harvest, but Enrico is in charge here and has a carefully devised strategy of what order to pick in, where to place the crates of grapes, and how to circulate the tractor to get them all without difficult maneuvers or needless backtracking. The heat is on, both for time and quality. We are urged to carefully select what we pick, snip off any burnt or rotten grapes, and leave the little immature clusters called *rasp' di San Martin* on the upper part of the vine.[1]

"Hey, be gentle with those grapes!" Enrico chides. "Don't throw them in the crate; *place* them in. And don't fill it too much, see? When we stack them we don't want to crush the grapes. Otherwise they'll start to ferment before we even get them back to the cantina!"

Next we head over to an old vineyard in Perno that Parusso has leased this year for the first time, and then up to a young one, in the Mosconi area on the other side of Monforte, that has just come into production. Then we go back down to the Bussia.

Parusso has three vineyards in the Bussia area: a portion of Rocche, the long, steep, narrow vineyard to the left of the road along the Castiglione-Monforte border; Fiurin (named after Bruna's oldest brother, Flavio, who owns it), and a swatch of my beloved Munie, a parcel right below Fabrizio's.

Though they are very close to one another and all contain the same nebbiolo vine, each of these vineyards is distinct and different; each has its own identity and microclimate, its unique combination of features—its personality. Some are old and crotchety with scaly wooden posts and sagging wires, while others are young and straight and clean. Some are so steep that you can barely keep your balance and have to clutch onto the vines to keep from rolling right down the hill, others are gently sloping. In the southeast-facing Rocche and Perno vineyards, you can watch the sun rise, gain momentum, and gracefully arch over the hill by high noon, leaving much of the vineyard in shadow, while in Le Munie it smacks you in the face mid-morning and beats you over the head all day long.

The special characteristics of a given vineyard—collected by the vine, transferred into its grapes, and reflected in the wine—are often referred to collectively as *terroir,* and trying to decipher them can bring a whole other dimension to drinking wine.

OF THE THREE primary factors that come to bear on the final product in a bottle of wine—the place in which the grapes grew *(terroir)*, the climatic conditions during the particular growing season (vintage), and the hand of man—the first is the most stable and fundamental. It is place more than anything else that determines the basic character of the fruit that comes from it; the whims of man and nature are secondary.

Depending on the region, the type of wine, and viticultural traditions, the critical factor of *terroir* may be subverted. In some cases of large-scale, commercial wine production, the effects of *terroir* are lessened substantially: the earth, pumped full of chemicals, acts as little more than a large petri dish in which the controlled cultivation of grapes takes place, with the forces of nature mitigated as much as possible by irrigation and other advanced technologies. One day perhaps such wine—defined in this case as a moderately alcoholic beverage based upon the flavor profile generally associated with the fermented juice of grapes—may be generated in a food-science laboratory, and the whole cumbersome process of grape growing and vinification dispensed with. The resulting product would be totally consistent, controllable, and duplicatable time and time again, leaving its creators to focus all their energies on marketing and promotion.

Such a beverage would probably be quite popular and commercially successful. But it would not be wine. Wine—*real* wine—is inexorably tied to a particular spot on the earth, as well as to a specific point in time, and is covered with the handprints of the people that made it. Without some intrinsic expression of or connection to place, wine is like a beautiful suit of clothes on a lifeless mannequin or, worse yet, an impostor trying to pass as something other than what it really is. While it might look pretty or be superficially enjoyable, such "false" wine is ultimately disappointing, much like sex without affection, religion without faith, or beautiful words without sincerity. Real wine is volatile and vulnerable and mutable, just like human beings, and it is precisely that which gives it uniqueness and value.

If, as I felt when I first joined Fabrizio in Le Munie, standing still in a vineyard offers a different experience from walking through one, working for long hours in a variety of different vineyards offers yet another. In time, you begin to feel a real relationship developing between you, the plants, and the particular bit of earth that is their home. Sometimes this relationship even begins to take on the characteristics of an intimate one: you wake up with the vines in the morning and tuck them in at dusk; you touch them and are in turn enveloped in their tall, leafy branches; you exchange bodily fluids (urine, sweat, grape juice) and drink together, they from deep inside the earth, you from a plastic bottle. You have long private conversations—well, no, I don't actually hear the vines talking, but my mind wanders freely while I work and they do seem to listen patiently, even sympathetically, to my internal ramblings. Occasionally they even seem to offer a silent commentary.

The vineyards may not talk, but they do express themselves—sometimes quite eloquently—through their grapes. (In this sense, one might think of great winemakers as translators or interpreters.) Like people at a cocktail party, grapes first say who they are (chardonnay, nebbiolo, cabernet, etc.), and one can often discern from their "accent" their country of origin. They also speak of the climatic conditions they experienced during their life on the vine, the age of the vines themselves, how they were cared for, and what happened to them after they were picked. But beyond that, what a wine really expresses is where it comes from; the things that make each vineyard unique and special; the specific combination of soil, slope or contour, altitude, and exposure: *terroir*.

The soil nurturing the vine contains a compact record of geological events that occurred thousands or even millions of years ago. Petrified sea creatures, marl, sand, limestone, and clay turn up in infinite combinations depending on how the earth shifted and changed over time. In a sense, soil itself is an agent of transformation; it takes whatever organic matter is introduced into it—leaves, branches, animals and insects, fertilizer—and breaks it down into a new composite material that the vine then transforms and transmits to its grapes.

Some people consider soil the most significant natural factor that affects *Vitis vinifera*. Others think it is the sun.

The sun supplies the energy necessary for photosynthesis (growth) to occur and for the grapes to ripen and develop sugar. It also warms the earth, helping it to process matter into minerals, creating a more conducive environment for the vine factory to operate. The sun is even more essential to plant life than it is to human life, for while humans can adapt themselves pretty quickly to extreme conditions, grapes do not grow in regions without sun or in places that are excessively hot or cold. Exactly how vines are positioned under the sun (known as exposition) makes a huge difference too.

Each vine represents a unique locus point or interface, bridging the sun and the earth, with secondary factors such as drainage, wind patterns, and length of time the ground has been cultivated all playing their part. These particular factors are most evident in a wine made from the fruit of a single vineyard. The grapes from each of Parusso's Bussia vineyards, for example, are picked, crushed, vinified, aged, and bottled separately in order to best show their individual characteristics.[1] This is the essence of *cru*.

Sometimes the grapes of a single vineyard are distinctive and complex enough to be bottled on their own. But single-vineyard bottling can, as Bartolo Mascarello liked to reiterate, also have its downside; sometimes the wine can be one-dimensional, reflective of a single personality with its particular strengths and weaknesses. In some cases, wines made from a blend of grapes from different vineyards can provide a more balanced reflection of a region, subzone, or commune,[2] though even such blends are usually made by harvesting and vinifying grapes from different vineyards separately and then combining them. Such wines might be thought of as a close-knit family of multiple generations, all of whom live in the same buildings as opposed to an isolated individual family member.

Each vineyard, whether its wine is bottled separately or blended with others, has its own personality.[3] The key is trying to recognize it.

Now, when I taste a wine from the Rocche vineyard, I *taste* the steepness, I *feel* the gravelly soil under my feet and the gravitational strain of trying to stand up straight, and I *see* the sun peeking up over the ridge to the east. When I taste one of Parusso's Mariondino Barolos, I sense the wide, gentle sweep and savor the open, forward fruitiness. And when I taste a wine from Le Munie, I feel the warmth on my head, the dark, dense, almost baked fruit, and the thick, sticky earth.

Much of this, you might say, is merely imagination or association. That might well be true, but much of the experience of tasting wine *is* associative. We have no way to understand a flavor other than by referring to other flavors or taste experiences we have had or could imagine having; how would you describe the flavor of an apple to someone who has never tasted one without referring to a pear or other similar type of fruit that they *have* experienced?

Of course, no one can spend time in every vineyard, but it really is not necessary. Much can be intuited from photos or descriptions of where the grapes came from, induced by the imagination, or informed by previous experiences and tastes.

But before you can even begin to taste *terroir,* you have to taste the wine.

28 · REALM OF THE SENSES

Wine awakens and refreshes the lurking passions of the mind as varnish does the colors which are sunk in a picture . . .

—ALEXANDER POPE

I taste [wine] for my profession and drink for pleasure, to forget, to remember, for nostalgia. Who knows . . . the reasons for drinking are many, no?

—NICO ORENGO, *Of Violets and Licorice*

MANY PEOPLE DRINK wine, but few actually *taste* it. For many of us, wine tasting is mysterious and intimidating, either because we haven't really tasted it or because we don't trust our impressions of what we tasted. Often, when we hear about someone else's experience—a so-called expert's opinion—we make ourselves believe we taste that too, just as we convince ourselves that we too truly admire the naked emperor's new clothes.

Of course, everyone is entitled to their opinion (indeed, when it comes to wine tasting, an opinion is not so much an entitlement as an inalienable right). But however experienced or knowledgeable others may be, it does not necessarily mean that I will (or should) taste the same thing they do, or that you will taste the same thing I do. Because tasting is a subjective experience, and each person experiences wine in their own way. Each of us has the potential to be our own very best sensory expert—provided you are not in a great hurry, have a modicum of sensitivity to your senses, and are not color blind, tone deaf, or palate free.

Wine tasting may well remain somewhat mysterious (as just about everything in the realm of the senses ultimately is), but it need not be intimidating. You don't need background information about the producer, the region, the soil, the exposition, the yield per square hectare, the grape, or whether the sun was shining at the moment it was picked. Such information *can* help enhance your sensory experience of the wine, but only if you have experienced it; otherwise it's just useless clutter. Everything you *really* need is right there in the glass in front of you: all you have to do is taste, really taste, again and again—it is never quite the same each time. But before you taste, take a moment to look.

The wine will likely be red or white, but that's pretty obvious. There's a wide range in each of these basic categories, a full spectrum, so let's focus in on the particulars.

Red ranges from near-black to pale pink and comes in a multitude of shades. Is this the dark red of a ripe black plum, the fire-engine red of a sour cherry, or the pale pinky red of watermelon? And what *quality* does the color have? Is it the matte of a weathered brick or the bright luminosity of an orange? The same is true of white wines, which range from rich gold to pale straw. Aside from color and quality, there is a tonal aspect as well: does the wine appear dense, thick, and syrupy like honey, or thin and transparent like air?

Think of color as a painter did in the Middle Ages when pigments, mixed by hand from natural ingredients, were never quite the same, never fixed and static, always imbued with something else. The color of wine is always different. It varies from grape to grape, wine to wine, and vintage to vintage (a warmer vintage will generally provide grapes with darker pigment). Color changes over time (an old red wine will often take on a somber orange-ish tinge while a young one will audaciously glow purple-red; an old white wine may either lose its luster and turn transparent or, if it contains a greater amount of residual sugar, take on a concentrated golden hue) and may even tell you something about how it was made (a thick, dark wine is likely to have spent more time macerating on the skins) or the grape it was made from (nebbiolo and pinot noir are generally lighter in pigment, barbera and cabernet are darker and thicker, while lagrein from the Alto Adige is positively black). Color also varies according to the situation in which you are looking at it; a wine observed in the flickering flame of a romantic candlelit dinner for two will appear radically different under the harsh light of day.

Look at it carefully, critically; hold it up to the light, tilt it back over a white napkin; ask your eye what it sees and listen to what it says without grasping for fancy adjectives. And if your eye is lazy, give it a slight push: "Yes, I know it is red, but what *kind* of red?"

Colors are nice to look at, just as it is nice to hear the sounds of birds chirping at daybreak or ropes banging against masts in a harbor. The colors of some wines go even further, practically pulling you into their world: a dark, dense, murky red lures like a Mark Rothko painting, a golden white showers you with warmth like taking a bath of butter and honey, while a pale rosé shimmers invitingly like sunlight filtered through gently rustling leaves on a summer day.

But while looking can be a worthwhile pause on a wine-tasting journey, it is not the destination; it is only a point of departure, an observation of appearance, an indication of what *might* follow. Looking at a wine is like a first encounter, sizing someone up, exchanging phone numbers, maybe a few little electric brushes of the hand. On the basis of that, you might even decide not to proceed any further. Maybe you can tell just by the look of things it's not your type: perhaps it's a dark red and you only do blonds; maybe it's got a freckly fizz and you desire a rich, creamy white. Or maybe there's some obvious problem: perhaps it has spent way too much time under a sunlamp (maderization), has an unhealthy pasty complexion (oxidation), or is much older than you first realized (throwing lots of sediment), and thus will never stand up to that spicy pasta you planned for a seductive dinner at home.

Chances are, however, you'll at least take a sniff, and with that sniff we cross over into intimate territory. Foreplay is about to begin.

Experiencing the bouquet, the "nose"—the *smell*—of a wine is a huge part of the experience of drinking it. In fact, smell and taste are the twin pillars of the wine experience and are practically inseparable. (You've probably noticed that when you've had a bad cold and a stuffed nose, your sense of taste is nearly nonexistent.) They occupy adjacent areas in the human body (back of mouth and nasal sinus), and both collect important sensory information to send up to the brain for processing.

The actual smell of something (or someone) is as personal and particular as the activity of smelling, and smells tell the smeller a lot about the special qualities of the thing being smelled. While you can see from a distance, you need to be pretty close to something to smell it, and this represents the first significant interaction. You inhale the aromas, actually take the scented air into your nose, and digest the evanescent impressions in your head. This is basically the same activity performed by every human being every other moment in order to sustain life. Breathing air that is scented or perfumed, however, lifts it above the level of mere survival, giving it a color, flavor, and additional dimension that greatly enhances this most basic of functions.

Smells are extremely complex, as complex as they are elusive. Sight and taste experiences have a kind of concrete objectivity to them: you see something in front of you much as the person next to you does; you put something into your mouth, it dissolves and releases its flavor. But smell is much more subjective and amorphous: it is invisible; it arrives on and disappears into thin air.

Smells say a lot about the thing that emits them, but they also say a lot about the person doing the smelling. No two people will perceive smells in exactly the same way. Someone who grew up on a farm is likely to have a whole different smell "vocabulary" from someone who grew up in a city. Some smells are easily recognizable, like the smell of sizzling bacon and steaming watery coffee outside a New York deli, while others are less obvious.

Bacon is one thing, but how do you describe the subtle smells emanating from a glass of wine?

This is where things start to get really complicated—and really fun. Here's an exercise, one you probably won't come across in most wine classes or magazines:

Close your eyes and forget about wine. Think of smell; focus on the middle of your face where smells begin to materialize. Think of three smells that are vivid to you, any smells, as long as they are pulsating with vitality and resonance. (I think of the musty, sparkly, "old-lady" scent of a jewel box my mother kept buried in a chest of drawers and would occasionally let me open and rummage through; the smell of hot, sticky tar melting on a dusty roof in the summer; the damp, earthy odor of the basement in the house I grew up in after a rain.) You may not ever smell any of these things in a wine (and even if you do, these descriptors might not be of much help in explaining it to someone else), but this exercise will help sensitize you to the vibrant world of fragrance.

Smell, you may realize, is often tinged with nostalgia. Smells stay with us with a special wordless potency and get filed away deep in our brains, becoming the trigger for memories of a whole host of other experiences. (I will never forget the smell of tall weeds and wildflowers along the train tracks during the summer of my first childhood romance.) Smells, though often hard to pinpoint, become part of who we are. And when you recall them or stumble across them in your present world, you can often be taken aback by the intensity of the déjà vu sensation. Most smell sensations have to do with a Proust-like remembrance of smells past: we perceive, compare, and categorize current experiences with those we have already had. And they, in turn, become added to our multilayered, ever-changing point of (scent) reference.

Okay. Open your eyes, bend over, and smell the liquid in the glass before you. What are the smells that come rising up to meet you? These passive free-run smells are generally referred to as the "aroma" of a wine. (It's kind of like smelling someone in passing, unnoticed. These smells tend to be rather muted and subtle, but they may be different from the more pronounced fragrances that follow, so why not enjoy them while you can?) Does the aroma jump out at you aggressively or hold back, coyly waiting for you to make the next move?

Go ahead and make it; swirl the glass vigorously (but not so vigorously that the wine goes flying out onto the white tablecloth—a definite faux pas). If the wine is an old one—as determined either by the distinct brown-orange tinge or a glance at the year on the bottle—swirl it gently; overagitating an old wine can hasten its downfall. Swirling a wineglass sometimes looks like an affectation, but it has a very practical purpose; it shakes up molecules in the wine and mixes them with air, which helps release both fragrances and flavors.[1]

Stick your nose into the glass and inhale deeply. Some people like to place their hand over the open part of the glass to intensify the smells and/or block out extraneous aromas, such as car exhaust, cigarette smoke, or perfume.

What do you smell? This can be quite a tricky question. Is it fruit? If so, what kind of fruit? Is it apples or berries or peaches or passion fruit? If it's apples, what kind? Green, crunchy, sour Granny Smith or ripe, juicy Red Delicious? Is it tart strawberries or dark musty blackberries? Or is it something else altogether: moist compost, burnt rubber, turpentine, cool metal, or the sappy scent of a lumberyard? Usually, it's not simply one or the other but rather an aggregation of numerous different smells. And it's always changing; smell the wine in your glass ten minutes from now or open another bottle of the same wine next year, and it may well be completely different.

Part of the difficulty is not so much smelling what you smell as describing it. Of course, unless you're a professional wine writer, you don't *have* to try to describe it or even think about how it smells; you can just sit back and drink. But if you do, not only may others be deprived of your insightful point of view, your own experience will be curtailed. Observing, exploring, analyzing, and articulating sensory experiences—even very briefly and just for ourselves—help to realize the overall experience in a deeper and more satisfying way.

It is difficult to describe something as intangible as a smell; the only tools we have are memories of other things we have smelled, an awkward, limited vocabulary, and our imagination. You can never describe a smell precisely, and that is how it should be, but it's nonetheless fun to try.

Here's another exercise: become a smell collector. Go on scent expeditions to interesting places where scents run rampant—foreign countries, the country, the city, back alleys, ethnic neighborhoods, county fairs, the zoo, professional kitchens during afternoon prep—but always be ready in the event an interesting smell just happens to cross your path. Actually, the smell doesn't necessarily need to be interesting or exotic, for it's not so much about the scent as the *scent*sation; it need only be one that generates a titillating sensory response. Imagine, for example, the clichéd all-American apple pie bubbling away in your oven: smells of sweet apple, caramelizing butter, flaky dough, and cinnamon fill the air (add, if you wish, a scoop of vanilla-bean ice cream, melted Cheddar cheese, or a splash of Calvados). Such an "everyday" experience could be a worthwhile entry.

Collect these scent-sensations and file them away somewhere. You don't need to write them down, or if you do, you don't need to show your notes to anyone. Use them as your own personal smell vocabulary or private reference library to help describe how wine smells. You may find that such scent collecting not only enhances your enjoyment and appreciation of wine but of the world around you as well.

Some wines show more to the nose than meets the palate. They are lovely, beguiling, and promising when you inhale them, but in the mouth they dissipate, fall apart immediately, show their true color—or lack of it. You feel let down, maybe even tricked. Sure, it's disappointing when things don't live up to your expectations, but that's how it is sometimes. At least you've had a worthwhile smell experience, perhaps even one you can add to your collection.

The best, of course, is when everything fits together; when an intriguing appearance leads to an intoxicating smell, and that in turn blossoms into a full-blown taste experience. You never know; some wines hold back in the nose only to explode in a veritable flavor bomb on the palate, while others that emit promising perfume fall flat once tasted. There's only one way to find out for sure. You've been sniffing around in intimate territory for some time now, the tension is rising, and there's no turning back: the moment has come to take the plunge and taste.

Though it may occasionally be disappointing, actually tasting wine is the climax of experiencing it, and we usually want to prolong it and extract as much pleasure from it as possible. So go ahead and take a sip (you may even want to take in a little air with the liquid to help maximize its aromatic qualities in the mouth), but don't swallow it right away. Savor it, caress it with your tongue, let it roll around in your mouth and over your palate. Let its flavors unfold themselves gradually. Before you even begin to think about how it tastes, think about how it feels.

Touch is the weakest of the four senses that pertains to the wine experience, but it still plays a part. Liquid has an identity distinct from all other things we might put into our mouths, and wine is a type of liquid that is heavier than water but lighter and less viscous than milk. Wine also has a texture and a temperature. Is it refreshingly cold and neutral, or hot like the Vin Brûlé we drink in Piemonte in the wintertime? Is it thin and strident or dense and velvety? Does it coat your palate like thick syrup or cut like liquid steel? The taste receptors in the mouth will often initially register things as physical sensations that then get translated into flavors. You can almost *feel* acidity (lemon) in the rear of each side of your mouth just above the jaw, an overabundance of sweetness (fruitiness) will almost make your teeth tingle (thankfully, this doesn't often happen with wine), and a sour or bitter sensation (tannin or bitter almond) will make your mouth pucker.

You may find yourself culling words to describe a wine from your experience of the world of touch. Well knit, tight, muscular, ethereal, silky, light-bodied, heavy, are primarily terms to help explain how a wine *feels* in your mouth rather than how it tastes. How it feels, along with how it looks and smells, is a big part of the overall impression and helps provide a context in which to approach the tricky world of flavor.

Think of all the things we use our mouths for: speaking, eating, smiling, pouting, singing. The very same lips we use to kiss, one of the most intimate of human interactions, also make a perfect cushion on which to rest a glass of wine as we pour it into our mouths.

Taste has developed along with the evolution of the human species. Initially, as animals and primitive *Homo sapiens*, we used taste to help us identify things that were bad (potentially deadly) and good (life sustaining). As we moved beyond the level of mere survival, we started to cultivate the ability to differentiate between various *types* of good, to hone our collective human taste buds and then to personalize them. For modern wine drinkers, taste has, thankfully, become less about preventing death than enjoying life.

Taste perception changes from era to era, culture to culture, and person to person, but we all have a remarkably sophisticated apparatus for deciphering taste and deriving pleasure from it.

At the base of the tongue is what is known as the palate, and this is where most of the action takes place. It's a good name; think of it as an artist's palette on which a multitude of colors are arranged, for the palate is made up of thousands of tiny nerve endings that receive flavor impressions in the form of electrical charges and send them up to the brain for processing. Actually, the rear of your mouth is kind of like NASA headquarters at liftoff; throngs of experts are sitting around back there ready and waiting to monitor each and every impulse and send it off for analysis.

Sounds complicated, but it is also surprisingly simple. There are only four basic types of tastes our palates can recognize: sweet, sour, bitter, and salty. (The Japanese have a fifth they call *umami,* a sort of savory fermented taste.) Salty doesn't really apply to wine, so that leaves only three. Only three! But, like computer language, these three basic flavor types can be arranged in infinite combinations, shades, and variations.

Flavors are often as hard to pin down as smells, and wine is not as straightforward or simple as bacon; each tastes different, and its flavors are often unique, complex, and evocative. Think you taste berries, plums, tar, licorice, shiny leather, musty basement? Guess what? They're not really there: what you are really drinking is just fermented grape juice. (Sometimes you can distinctly taste the toasty oak tannins in wines aged in new *barriques* or infused with wood chips, though if the wood is not well integrated with the other flavors, this is generally considered a flaw.) Because flavors in wine are practically all secondary, our personal experience, imagination, and memory are automatically called into play.

How do you accurately register a taste impression for yourself, much less describe it to someone else? And how do you differentiate *this* wine from others you have tasted before? But we're getting ahead. You've looked at it, twirled it, smelled it, and rolled it around in your mouth, so there's nothing left to do but *taste*.

Take a sip of wine and hold it in your mouth for a while before swallowing it. What do you taste? What did you taste when it first entered your mouth, and what do you taste now? What do you taste after you swallow it?

The taste experience is not static and one-dimensional. It develops over time, has a beginning, middle, and end, a trajectory like the arc of an arrow flying through the air. Sometimes it even goes darting off in many different directions like one of those fancy fireworks.

The first instant when the wine hits your palate is called the *attack*. It can be quick and aggressive or slow and subtle. Last is the *finish*, as the wine slips down your throat like the sun setting behind La Morra, and then the *aftertaste*, the fluorescent pink clouds left in its wake. Taste proper occurs in between the attack and the finish, and this is usually the longest and most prominent part of the experience. As the flavors change in the mouth, tasters will sometimes refer to the "middle" or "end" of the palate to describe how a wine evolves prior to swallowing it.

Now close your eyes and taste, *really* taste. Feel the wine in your mouth and let its flavors blossom on your palate. Try to describe it to the person sitting next to you, and listen to what he or she says about the same wine. It may be similar or it may be quite different; you may even hear something new that will enhance your own experience. But don't let it sway you; you yourself are an expert, and when it comes to *your* taste, you are the only one who really counts.

Sometimes it's fun to go through a thorough critical examination of a wine, experience it on each and every level, and flex your bulging muscles of sensory perception (Pair a wine with food and a whole other dimension of analytical complexity arises). Other times you might just want to drink without analyzing or critiquing. Either approach is fine. The important thing is to listen to your senses, hear and trust what they say. And taste again and again.

Actually, there *is* a sixth sense that plays a part in the wine-tasting experience, and it may be the most elusive of all. Call it the Sense of Place. Wine is the composite product and expression of where it came from, of its *terroir*, so a sense of that place, whether based on firsthand experience, research, intuition, or imagination, can greatly enhance your enjoyment and appreciation of drinking it.

Working in multiple vineyards, I become intimately familiar with their particularities. I love the distinctiveness and diversity of the different vineyards I work in, and I like to try and *taste* how the outward differences might be reflected in the flavor of the fruit. But the more vineyards I work in, the more I feel a special bond to Fabrizio's little section of Le Munie, my first and longest, closest and deepest relationship with the vine. This is *my* vineyard—not in the literal sense, of course—but the special affinity I feel does give me a sort of proprietary interest.

It goes without saying that I'll be there for the harvest; for me, it's the culmination of the past eight months. From the very first pruning in January—the removing of dead wood, a lone remaining branch, a fresh start—through the bending and tying-down of *tòrse*, the resurrection and rebirth of spring, and the sweaty battles of summer, my experience with the vines in this vineyard has been not only a reflection of my present life but a vehicle for it, an agent of change.

A year ago everything seemed in ruins. Then, once the debris had been removed, it seemed empty, like the fenced-in hole where the Trade Center had once stood. I felt lost and confused. Now things are different: I am living in Italy; I am with Ivana and working outdoors in the hills of the Langhe among the vines. The "scars" may still be there, the transformation not yet complete, but something has changed; something in my life seems to be coming to fruition, just like the fat purple grapes hanging in Le Munie.

We never talk about it, but I think Fabrizio understands this in some way, my attraction to the vineyard and the significance of my "lending a hand." I'm sure he knows I will be there to pick the grapes. But out of politeness—*per educazione,* as the Italians say—he asks anyway: "*Mi dai una mano a vendemmiare, no?*" ("You'll lend a hand with the harvest, won't you?")

"*Si, certo,*" I answer aloud, adding to myself, "Wouldn't miss it for the world."

29 · STRIPPED BARE

WE GATHER AT the top of the Munie road at 7:45 A.M. on the last Saturday of September. It is a clear cool day, but the autumn light has a warm lemon-custardy glow that seems to cover the vineyard with a thick, transparent, luminous blanket. You can almost feel the grapes tingle in anticipation of being plucked from the vine, released from their tethers, and setting off on a whole new journey.

Assembled are Fabrizio, jubilant beneath his cool, expressionless demeanor; Franco Conterno, flanked by his eldest son, Andrea, and his Macedonian worker; and Sebastiano, Fabrizio's father-in-law. Marilena is there too, and she has brought along baby Pietro to witness his father's first *vendemmia*.

"You *did* bring your camera, right?" is the first thing Marilena says to me. Without saying a word, I jump in a car, rush home, and am back shooting within five minutes.

At eight we begin. Of course, I know what to do—I have done it
many times now—but it's different to be doing it *here*. It feels quite odd
at first to have other people in "my" vineyard, where Fabrizio and I
have spent so many solitary hours. Then, just as I am getting into a harvest groove, it is done. There are six of us working through the little
vineyard and before noon, in less than four hours, we have gotten
everything. I take a picture of Marilena and Fabrizio (who has finally
allowed himself to smile) in front of his little orange tractor piled with
red crates of purple grapes. Then all the crates are loaded into Franco's
truck and driven to Castiglione to be weighed.

The public scale, a long, flat metal rectangle set flush into the asphalt with a little glass booth behind it, is located in a parking lot right
by the Alba-Monforte road, across from my apartment. Just about
every town around here has one. It sits there all year long, but is only
really used during harvesttime. We buy tokens at the Coccinella bar;
one gives access to the booth and another activates the scale. Once the
truck is driven onto the scale, levers are used to gauge the weight,
which is embossed on a strip of paper. Then we drive to Franco's cantina, dump the grapes into holding containers, put the empty crates
back into the truck, and return to the scale to weigh everything again.
This new measurement is embossed on the same strip. The difference
between the two numbers is the net weight of the grapes—2,600 kilos
in this case (a little over 5,700 pounds)—and the strip of paper becomes
the official record of the harvest of that vineyard. Fabrizio is a bit surprised. He knew the vines were old and the season not especially boun-
tiful, but he still expected a bit more. His disappointment, however,
passes quickly; this morning, contentment rules.

After the weighing is finished, Franco hauls the empty crates back
to his cantina, and we head to the Bussia for a celebratory lunch. In the
Mascarello courtyard, a long makeshift table has been set up and it's
covered with a collage of paper tablecloths with rocks on top to keep
them from blowing away. Pietro and Marilena are already there, as are
Sebastiano and his wife and Ivana's mother. (Ivana is still at work.)
Argo is there too, of course, and seems especially excited, whirling
around and around in circles trying to bite his tail.

Marilena's father, Sebastiano, is a little guy with a deeply lined face and a permanent indentation circling his head from the narrow-brimmed straw hat he always has on. He wears heavy flannel shirts and thick wool pants, with patches, that ride high up on his short, compact frame, even in the summer. Sebastiano (or Bastian, as he is known in Piemontese) is a farmer. He lives with his wife and two unmarried sons on the far side of La Morra toward Cherasco, just outside the Barolo boundary. He has about six hectares, which is planted to vine (mostly dolcetto) plus some grain and vegetables. He used to have over forty head of cattle too, but has let them all go except for five, which he keeps in a stable adjacent to the house. He's getting too old to manage everything, and none of his children has expressed much interest in taking over.

I had met Sebastiano once before, at the recovery center after Ivana's mother's operation, but we had barely exchanged a word. Now, Fabrizio is off washing up, the women are busy indoors, and I am banished out to the courtyard with Sebastiano. As I slide into a plastic chair, Bastian approaches from the other side and sits down. Once he gets settled, he pours us each a glass of wine, raises a deeply grained hand with dark bands under the fingernails, and opens his mouth. After a moment or two a breathy stream of words comes tumbling out—and I don't understand a single one of them.

"*Come?*" I ask. "What did you say?"

He happily repeats his statement, but it doesn't help.

Not knowing what else to do, I let out an emphatic "*Unh!*" of assent to whatever it was he said. This response seems quite satisfactory because Bastian makes another unintelligible statement and then breaks into wispy heaves of laughter. I can't help laughing too. We go on like this for a good fifteen minutes, and by the end you'd think we were two old buddies who hadn't seen each other for a while catching up. The only word I did understand for sure was *ducette* (dolcetto); we had emptied the better part of a bottle during our "conversation."

Gradually, everyone else took their places around us.

Lunch was simple and satisfying. There were the usual things: salami (cooked and *crudo*) and prosciutto, pinched ravioli with meat sauce, and a roast. Nothing special. But what was truly memorable was the collective feeling of satisfaction and accomplishment we all shared:

the first harvest of the first season of Fabrizio's working the vineyard was completed, my first season in Piemonte finished. A cycle had come to an end.

Having finished her shift at the hotel, Ivana arrives as we are finishing our meal, squeezes in next to me, and wolfs down the tepid food that had been set aside for her. Her mother then brings out the classic chocolate pudding called Bonèt. Ivana helps herself to a healthy portion, and then a second before I have even made a dent in my first. *"Mmmm, che buono!"* she says, smacking her lips with obvious and unadulterated delight (Bonèt is her favorite dessert).

Someone brings out coffee, and a bottle of Barolo Chinato is placed on the men's end of the table. Then, just as the afternoon sun begins to tilt over the ridge above the village of Barolo, Fabrizio says, "So, shall we get back to work?"

It is more of an announcement than a question.

The women don't move a muscle—their tasks are finished for the time being.

"Uh, sì," I say. *Work? What work? Haven't we just finished harvesting the vineyard?* Bastian makes a wispy groan of assent as he drains the wine from his glass and gets up from the table.

We return to the now grapeless vineyard and, clippers in hand, position ourselves at the head of the first three rows. I take the third row and take my cues from Fabrizio above at the second. We clip the vine branches at the base of the bent branch, gather them all together, and pull—pull from the gut, pull with the weight of our entire bodies. It's not easy. Sometimes the vines don't budge and need to be divided in half to get them to come down; sometimes they are too tall and too intertwined to get off whole and need to be cut in half lengthwise. When, one way or another, they finally do succumb, they are laid in as compact a mass as possible down the middle of the row so the tractor can pass later to pick them up.

"Couldn't we do this later?" I think to myself. "Like some other day?" I want to savor the joy of harvest, this climax of the first season in Le Munie, keep it separate in my mind from this other brutish, post-climactic activity. It seems as though, having plucked all the fruit from the vines, we have returned now to pillage and destroy. The work is reminiscent of the pruning we did the very first time I came to this vine-yard, except that now the vines are still alive and supple; the leaves are still green, the tendrils still firm and taut. And there is no pause to choose a branch, count buds, or ponder the possibilities: just cut away and rip 'em down, everything, right down to the thick gnarly vine. It would seem that there is no technique here, just savage cutting. But why then do Fabrizio and Sebastiano move steadily ahead of me?

Sebastiano, the oldest of our trio by far, is always the first to finish a row. He doesn't seem to exert himself; he makes a few choice cuts and positions himself with one foot forward as he grabs the branches to-gether like a big bouquet. Then, mouth puckered as if to whistle and watery eyes glinting, he gives a sharp tug and the branches seem to yield easily to his wrinkled, callused hands. He lays them gently in the alley before moving methodically on to the next vine and the next. He finishes his row before I am even halfway down mine, then turns to give a wave of support and a raspy chuckle before heading down the hill to his next row.

We get through about a third of the vineyard before the sun, merci-fully, sets. The next day, Sunday, we go back to finish up, after which the vineyard is reduced to a dusty field of naked woody stumps. Later in the week we return to take down the wires and remove the old wooden posts.

The following weekend, Franco is back with his tractor, the yellow soot-spewing beast dragging a heavy chain from each side. He maneu-vers it slowly down the middle of each row while Fabrizio and I get down on our hands and knees to scrape the dusty earth away from the base of each vine stump with our fingers. We wrap the chain as low down as pos-sible beneath the bulging knuckle of the vine; then, as the tractor moves ineluctably down the row, the slack tightens and the stump comes flying up out of the ground. We follow behind, loosening the chain, tossing the vine stumps into a pile, and securing the chain onto the next stump.

A few days later we cart the vine stumps over to the field where the hazelnut trees used to be. They will be dried out and later used to feed the wood-burning *stufa* in the Mascarello kitchen.

The vineyard is empty now, but there is still one thing left to do. I get Fabrizio's call one Saturday morning in late October.

"Hey, can you come give me a hand this afternoon?"

"Sure," I say. I don't ask what he needs a hand for; I know it can be only one thing.

Tiziana Parusso had raised the subject some time before.

"What's he going to do with the well? Tell him not to take it out. I wouldn't take it out. Especially not after the drought we had this season. You never know. They almost rationed the water this year, *sai*? And all that stuff they put in it now, all the chemicals; we prefer the natural water that comes out of the ground, and right from the same ground as the vines too. Do you know how long it takes to get permission to dig a new well? If it were mine, I'd leave it."

I had raised these concerns (without saying where they came from) to Fabrizio on a number of occasions, but to no effect.

The old well was a wide-open cylinder of brick rising about three feet above the ground and going down thirty feet or so into the cold dark earth, mushrooming out at the bottom where there was an ever-present pool of murky water. It was impossible to see all the way down, but you could throw a pebble in and, after a second or two, hear a soft plunk echoing up to the surface. For nearly a year we had been working around it. It sat right in the middle of the second and third rows from the top, making it impossible for the tractor to pass. Even on foot you had to squeeze between the well and the vines to get by and take care not to fall down into its dark watery depths.

Now, without the vines around it, the well stuck up out of the naked vineyard like a big pimple on an adolescent's otherwise pristine skin.

I wasn't really in a hurry to get there that afternoon, so I decided to walk through the vineyards rather than take the asphalt road.

I set out on a dirt lane called Pugnane just steps outside my door and turned left onto a path that passes through the vineyards. I walked along the upper edge of Villero, and there below, immersed in the vines, I saw Bruno, a neighbor of mine from Castiglione and one of the *Radio Trave* regulars. I waved. He looked up at me but didn't seem to know who I was, so I took off my hat to expose my face and went down into the vineyard to say hi.

"Oh, it's you. I didn't recognize you," he drawled out. Bruno had a stroke a few years ago and speaks in a slow, staccato grumble, as if squeezing out the words with difficulty. He used to be a bus driver but is retired now and spends most of his time walking from his house at the foot of the village up to the piazza and back.

Wearing a green pullover sweater and creased pants with black leather shoes, Bruno looked quite out of place in the chunky, dusty gray earth among the vines, as if he had taken a wrong turn on the way home from Sunday Mass and wound up here.

"Yes, it's me," I said. "Beautiful afternoon, no?"

He just continued gazing out over the green expanse with cicadas clicking in the distance.

"How are you?" I tried again.

"*Eh beh,* how should I be? I'm only half a man," he spit out as if disgusted.

"What do you mean?"

"I can't do anything anymore. I used to drive a bus full of people around to visit the big cantinas. And drive the kids to school. Whenever I had a free moment, I would come here," he said, slowly turning his head and the upper part of his stiff torso to indicate the downward sweep of vineyard. "I came here to work. But I can't drive anything anymore."

After a long pause he went on: "This was mine. I put these vines in here and took care of them. Any chance I had, I would always come here. I loved this place; I felt good here. And useful."

There was another long pause during which I could hear a lone bird twittering in the forest below and an occasional car whooshing by on the road far above.

"We took out the fruit trees down there by the bottom and put more vines in years ago. More vines. Look, it needs to be hoed here, and it's time to start pruning. But I can't do anything anymore. I'm only half a man. Now my son-in-law takes care of it. He wants to change the poles; they're old, like me. He wants to put in new ones, straight and strong. Sure, it's a good idea; they should be replaced, these poles. But I can't help. I can't do anything." As he said this, he placed both hands on a nearby headpost, grasped it, and lowered his hulking body forward until his forehead touched the wooden post. Tears appeared in his squinty eyes.

"Listen," I said after a while. "You're not half a man, you're a whole man. It's just that there are different times to do different things. You've done your part here and you should be proud. Now you can let someone else worry about it. It's a beautiful vineyard, really *un bel pezzo*, and you made it. It wouldn't even be here if not for you. And now you can pass it on to your children, no? Am I not right?"

"*Sì*, I suppose that's right. But all the same, I miss it. I used to love to come and spend time here whenever I could. I put my whole self into this place, everything I had."

"Yes, I know. And you should be proud. But what you put in here has not vanished; it is still here and always will be, even if they change the posts. It doesn't matter; it's in the earth. And you can come here whenever you like. You live close by. Just come and *be* here—and let someone else worry about the work."

This dialogue went on for quite some time, because he spoke very slowly and with frequent gaps of painful silence. I was touched by the chinks showing in his normally impervious coat of armor, by his affection for his vineyard and longing for the past. I didn't want to just leave him there alone. But it was also getting late and I had visions of Fabrizio falling down the well with no one around who could call for help.

"Listen," I said. "I have to get going. I have to go and help Fabrizio over in Le Munie. But think about what I said. You know it's true: you're a whole man, it's just different. Time passes and things change, just like the seasons, and there's nothing we can do about it. Let your son-in-law do the work now; you can just come and enjoy it. You deserve it. You should be proud. Everything you put in here is still here and always will be."

"*Beh*, yes, I suppose you're right." Then after a pause, "Mascarello? I knew his father, Francesco. *Un bel pezzo*, that one. *Roba buona*, good stuff, just like here. Yes, I think I'll be moving along too. *Ciao*."

I continued up to and along the path through Villero to Mariondino and over to Le Munie. As I walked, I turned frequently to watch Bruno climb laboriously up out of his vineyard and disappear into the trees on his way back home.

By the time I got to Le Munie, Fabrizio had already removed the circular wall of thick, uneven terra-cotta bricks that had stood above the ground. I made a few last-ditch attempts: "So, are you sure you want to take this out?" . . . "Hey, you never know when a free supply of natural water might come in handy, no?" . . . "Maybe we could just cap it and throw some dirt over the top, just in case?" But my questions fell on deaf ears: the well was coming out, and that was that.

Fabrizio worked his way down into the hole, pulling out bricks, and tossing up them to the surface. With each one that came up, I felt a pang of regret.

"How old is this well?" I asked.

"Oh, who knows? It was dug before my father's time, maybe by my grandfather or great-grandfather."

That, I realized, could put it easily at over a hundred years old.

This well, made by hand, was an impressive work of craftsmanship in its own right; the deep hole perfectly round, the handmade bricks laid without cement. Who knows how long it took to build? It surely required ingenuity and skill, not to mention sweat, perhaps even blood. And now we were undoing all that work.

It was the only task in the vineyard I regretted, and I still think it was a mistake. But there was nothing for me to do but load the old bricks onto the tractor bed. When the last brick was removed, Fabrizio pulled himself up out of the hole by a long rope attached to the tractor and we carted the bricks back to his mother's house in the Bussia for some future use. Several days later, steam shovels came to fill in the hole where the well used to be. And then it was done: the last remnant of the old Le Munie was gone. The cycle of this old vineyard had come to an end, at least for a while. And a new one was about to begin.

With the nebbiolo harvest complete, the season was officially over. The final verdict: good, perhaps very good.

Despite a nail-biting roller-coaster ride of a summer, in the end the vines did well and people were basically happy. Quantity was down about 30 percent—not necessarily a bad thing with the glut of wine on the market. And while some complained about a lack of pigment in the nebbiolo, it was more than compensated for by the beautiful (and powerfully alcoholic) dolcetto and barbera. Because of the extreme heat, the grapes had ripened very early and very quickly. Those who could took a chance and left them on the vine as long as possible, but they still lacked the maturity of slow ripening over the extended length of a normal season. Chances are 2003 will not be a stellar year for long cellaring, but that is not necessarily a bad thing either under the circumstances; this vintage is likely to produce good to excellent wines to be drunk young.

Likely—but even at this point nothing is certain. People now know about the grapes, but the actual quality of the wine they will make depends on what happens in the cantina over the next few weeks and the next few years (three years, in the case of Barolo). Most winemakers, as cautious and as superstitious as the farmers, are reluctant to make predictions: "Looks pretty good," they say. "But we'll have to see what happens."

30 · INTO THE *CANTINA*

AFTER THE FINAL weekend with Fabrizio in Le Munie, I returned to help Parusso pick the last of the nebbiolo in the Bussia. Before long, that too was finished, and I rode back to the cantina standing on the tailgate of the last grape-laden tractor, proudly watching the dips and curves of the Langhe pass by while my sticky hands grasped on to the rickety mountain of crates. We pulled up the driveway into the courtyard like weary, victorious soldiers returning from a long campaign: the harvest was over, our mission accomplished, our booty brought home! But we wasted no time in frivolous celebration; there was work to do.

We jumped down and immediately began stacking the bulging crates onto wooden pallets.

*   *   *

There are two distinct phases of wine making; growing grapes and actually making wine out of them. These are separate activities requiring different skills and sensitivities and for this reason are often handled by different people. Wine making is much more capital-intensive than grape growing and requires more technical expertise, while tending vines and growing grapes is extremely labor-intensive. Historically, there were many growers but few winemakers (that is, those making wine for other than their own personal consumption) and most of them were under the auspices of the nobility or the church. Later, *negoçiants* or co-operatives emerged, well-equipped commercial enterprises which collected grapes from many small growers for vinification and distribution. Sometimes—more now than ever before—grower and winemaker are one and the same.

Because wine grapes are grown to be made into wine, and wine is only as good as the grapes it is made from, these two phases, while distinct, are fundamentally and intimately connected. And the moment the freshly picked grapes enter the cantina is the fulcrum between them.

The grape transforms and encapsulates the essence of a place, much as wine, through fermentation, vinification, and aging, transforms and expresses the unique qualities of the grape. The grape is tied to a specific place and a specific time (vintage) with its certain set of climatic factors, while wine is like the afterlife of the grape, immortalizing and preserving it right up until it is drunk. But this is not a static embalming: what takes place in the winery actually gives a whole new life to the grape by making it into something else altogether. The journey from fruit to wine is itself a transformation almost as mysterious and magical as the biblical transubstantiation of wine into blood. The grapes we had just finished picking and were now busily unloading from the tractor were about to set off on this next passage. And I was going to get to tag along.

"*Vai! Vai! Forza!*" shouted Francesco over the constant mechanical roar.

I was up on a platform made of wooden pallets, lifting crate after crate of grapes off another pallet suspended on a forklift, and dumping them into a trough.[1] The trough in turn brought them up a conveyor belt to the destemmer and into the cantina.

In response to his imprecations I immediately switched into high gear.

"Oh, *piano, piano!* Let's not overdo it, eh!" he shouted again, a little less urgently. "We want the belt to be full but not *too* full, no? Otherwise they get crushed on the way in and the juice winds up on the floor. (But at the rate you were going before, we'd be here all night!) Not too fast but not too slow either, okay?"

This machine, positioned right in the middle of the huge open double-doorway, is the grape's portal into the modern cantina. A conveyor belt carries the grapes up and drops them into the destemmer's perforated cylindrical drum. As it spins quickly, it separates stem from grape by centrifugal force. The stems remain in the drum (periodically the drum is emptied, the stems are shoveled out a hole in the cantina wall and allowed to dry; they are later burnt and the ashes used as vineyard fertilizer) while the grapes—now somewhat broken—and their juice are pumped into the big horizontal rotary maceration tanks well inside the winery.

In the midst of all this noise and activity Signora Rita, matriarch of the *azienda* and grandmother *(nonna)* of Marco's and Tiziana's children, comes marching up with a huge pot.

"Hey, give me some juice! It's time to make Cugná." This is *her* harvest ritual.

Francesco and Enrico drop their other chores, take the pot, and go inside to fill it with unfermented grape juice, which is called "must." They bring it back nearly full and Rita has them set it over a low burner in the garage. She cooks the must slowly for hours, stirring occasionally with a long wooden spoon, until it is reduced to a thick syrup. Halfway through the process she adds quartered Madernassa pears, fresh figs, and hazelnuts. Sometimes she needs to add some additional sugar too but this year it is not necessary; the grapes have plenty on their own. The next day, once the Cugná has cooled, she ladles it into sterilized glass jars, which are dated and stored away in the pantry for use throughout the year as an accompaniment to boiled meat and aged cheeses.

A generation or so ago, people hauled their grapes back to the cantina in a horse-drawn cart, dumped them into a big wooden vat or press, and stomped on them. The juice fermented by itself pretty much as it always had, starting spontaneously and stopping when it was done. The wine was then stored in large wooden casks called *botti* until bottled; often it wasn't bottled at all but sold *sfuso* in demijohns or barrels. Though the process is fundamentally the same, these days wine making is a bit more complicated and controlled.

These big horizontal temperature-controlled rotary fermentation tanks are cutting-edge technology. Inside, the grape solids (mostly skins) rise to the surface forming a thick cap, while the seeds and other heavier particles sink to the bottom. Because much of the tannin and nearly all of the color come from the skins, it is desirable to periodically break the cap and submerge it in the must to extract color and flavor and soft tannins. This was originally accomplished by someone plunging himself into the open vat while hanging onto a bar overhead. Nowadays that is no longer necessary. Modern tanks have a long paddle inside, which can be programmed to revolve slowly, gently mixing the skins with the juice. Other modern fermentation vats are vertical and have a plunger that breaks up the cap or a screen that submerges it under the surface of the must. In old-fashioned vats of cement or stainless steel without fancy paddles or plungers, a technique called "pumping over" is used. To pump over, the liquid is drawn out through a nozzle in the bottom of the vat and pumped through a hose up to the top where it sprays over and filters through the cap. This is done several times a day. However it is done, this remixing and steeping of the solids in the liquids during fermentation is called maceration. Maceration times vary: a big red wine may spend weeks macerating, a light red a few days, a rosé hours; a simple white wine, on the other hand, may be sent directly to the press and fermented afterward.

The destemmer grinds on tirelessly with only an occasional pause to empty out the stems while we take turns up on the platform. In a single day all the plastic crates have been emptied and the wooden pallets stacked away. After the last crate has been dumped into the trough and the grapes have run through the destemmer and the pump, the machine is turned off and a hush falls over the cantina. The machine is then disassembled, thoroughly hosed down with hot water, and stowed away in a corner of the winery; its work has been done and it won't be needed again until next year.

While we deal with the destemmer, Marco goes up to the fermentation tank, opens the glass door of a fancy box on the front, and begins tinkering with the computerized controls. He sets the internal temperature for 7 degrees Celsius (about 44 degrees Fahrenheit), and a moment later, the actual temperature of 14 degrees Celsius appears on the LED display. Cooling elements in between the double walls of the tank will quickly bring the temperature down to the desired level. Next he sets a program for the paddle rotation: clockwise, one revolution per hour for eight hours followed by a three-hour break. The grapes are then left there (under close scrutiny, of course) to macerate while we continue with our postharvest cleaning.

About a week later, we are ready to take the next important step.

Francesco calculates the amount of liquid now securely contained within the horizontal tank and prepares a fragrant, milky mixture of yeast in a white plastic bucket. He leans a ladder up against the front of the tank, carefully climbs up one-handed, grasping the bucket handle in the other, and pours the contents in through an open hatch on top.

Marco sets the thermostat to 30 degrees Celsius (about 85 degrees Fahrenheit) and before long, the digital readings start to rise. (He calls this rapid increase in temperature following cold maceration "heat extraction.") At the end of the day, as the computerized reading reaches 27 degrees Celsius, he sets the desired temperature for 20 degrees (about 70 degrees Fahrenheit). Even without looking at the computerized readout, it is obvious something is going on; the outside walls of the tank are warm to the touch, there is a sweet yeasty aroma wafting out and occasional echoes of bubbles bursting within. Fermentation has begun.

Fermentation is a chemical process in which yeast, a kind of benevolent bacteria, interacts with sugar and creates alcohol. A similar transformation occurs in the creation of many other substances, such as beer, bread, cheese, and soy sauce. But the fermentation of grapes is special.

Grapes are naturally coated with a thin layer of white powdery yeast, which forms during the later stages of growth after the grapes have begun to turn color. While still on the vine, it acts as a protective coating for the fruit, helping to deter other kinds of bacteria from forming (during leaf cleaning and cluster untangling, the grapes must be handled as little as possible so as not to remove this yeast layer). When the grapes are crushed, the yeast on the skins turns from protector to protagonist, interacting with the sugar in the juice and pulp and initiating a chemical reaction. In addition to that present on the skins, there are also millions of invisible yeast cells floating throughout the air in wine-producing areas, which at harvesttime heed the sweet call of freshly crushed grapes.

Fermentation will occur spontaneously on its own but you can never be sure when or exactly which lucky yeasts will make it happen. For this reason, most modern winemakers prefer to control this critical transformation as much as possible by choosing the yeasts that are added to the tank to induce fermentation. When the must is collected and all the other factors deemed desirable, a measured amount of selected yeast is mixed with warm water and added to the must. This is known as "inoculation" and that's what Francesco was doing up on the ladder.

However it is initiated, yeast acts as a catalyst for the chemical transformation of the sugar in the grapes into alcohol, turning grape juice into wine. A "dry" wine is one in which nearly all of the available sugar has been converted to alcohol; a "sweet" one, often made from late-harvested grapes with an extremely high level of sugar and sometimes infected with a "noble" (that is, desirable) mold called botrytis, is one in which there is a higher degree of sugar left over after fermentation is complete. This leftover sugar in the wine is referred to as "residual."

Fermentation—like many other types of transformations—is a dynamic process in which energy is created and expelled. The temperature in the tank rises, gas is created, and you can hear the bubbles bursting and gurgling inside the tank and echoing off the thick stainless-steel walls. If you stick your face over the hatch at the top (as I had to do on numerous occasions in order to monitor the position of the paddle), you get a rush of potent fumes that is almost immobilizing. As the wine ferments, the yeast cells that effected the transformation gradually die and sink to the bottom, forming a thick, muddy sediment. Fermentation—depending on the grape, the wine, and the year—usually lasts from one to two weeks, though it can go longer.

After a few days, a small amount of turbid liquid is extracted through a nozzle in front of the tank, put into an airtight vial, marked with the date, the time it was extracted, and the vineyard and tank it came from. This sample, along with samples from other tanks, is then sent off to a lab in Serralunga for analysis, and the results are made into charts. To me these charts are little more than watercolor blotches with incomprehensible notations, but to Marco and his associates they demonstrate exactly where, chemically speaking, each cask of liquid is along the way to becoming wine.

About six days after fermentation began, Marco gets a chart back from the lab and scowls:

"Cazzo! The yeasts are going crazy eating up all that sugar! It's going way too fast. *Non va bene. Ecco!"* He pokes a button to bring the temperature down another six degrees.

Another feature of these modern fermentation tanks is that the winemaker can prolong fermentation by gently lowering the temperature (cold fermentation helps retain the aromatic qualities of the wine) or raise it to speed the process along. In a hot year like this one when the sugar content of the grapes is so high, most winemakers try to slow the process down as much as possible (in the past they sometimes heaped ice around fermenting vats or put plastic bags of ice directly into the fermenting must); in a cold year, raising the temperature in the tank can create a more favorable environment for the yeast to do its work, thus avoiding the not uncommon problem of an overly long or incomplete fermentation.[2]

Due to the exceptional heat of 2003, many winemakers—especially those with less sophisticated equipment—are experiencing difficulty reining in fermentation; the intensely sweet, superripe grapes want to throw themselves into a wild abandon of chemical activity. The danger is that the resulting wine may turn out to be fat, fruity, and overly alcoholic as well as dull, superficial, and, ultimately, unstable.

Subsequent analysis gets no more than cursory inspection. Then finally, after more than ten days of waiting and watching, temperature-adjusting and paddle-rotating, the contents of the manila envelope just back from the lab elicit a notable reaction: what we have been waiting for has happened; the amount of sugar has significantly lessened while the degree of alcohol has risen and then leveled off. Marco calls his enologist on the phone; the facts are confirmed and the decision is made: *"Okay. Ci siamo!"* The miracle that has taken place has been verified by science; grape juice has become wine, and Barolo Bussia 2003 is about to be born.

When the first fermentation (also known as the principal or alcoholic fermentation) is nearly complete, the temperature in the tank is decisively lowered to end the fermentation process and help stabilize the new wine. (At this point, some active "live" yeast cells that haven't found any available sugar to impregnate are still present in the wine.)

"Here. Want to taste?" Marco has filled a glass from the nozzle and is holding it out to me, an impish smile on his face. "Uh, yes. Yes, I would!" I say, trying to sound convincing. But the contents of the glass don't look very appealing: the liquid is cloudy, it has the blotchy purpleness of someone who has taken a bad beating and there's stuff floating in it. Fruit flies are flying all around the glass and, as I take it by the stem and tilt it back to my mouth, I notice that a few are lying listlessly on the surface. Bad as it looks, it tastes even worse: it is tough and astringent, highly acidic, obnoxiously fizzy, intensely alcoholic, and emitting sulfuric fumes. As if that weren't bad enough, when I hand the glass back to Marco, I realize that there's one fewer fruit fly floating on the top.

"Buono, eh?" Marco says, not making a move to take the proffered glass out of my hand. "There's great fruit there, that's for sure," he continues. "And the alcohol's off the charts. The tannin's good too. It could use a bit more color, but what else is new; we'll leave it there overnight and let it get a bit darker."

I make myself try it again. It's neither easy nor pleasant to taste brand-new wine from cask; technically speaking, this *is* wine, but it's rough and raw, certainly not something you want to sit down and have with dinner. This is still very much a work in progress. Being able to appreciate a wine in this state is not so much a question of taste as of knowledge and experience, of being able to extrapolate from this harsh, chaotic mass of undeveloped characteristics some idea of what it might eventually become. Maybe he's right, I think to myself as I digest my second taste. The acid and alcohol are still nearly overwhelming but my palate is a bit less shocked than it was the first time. The liquid is as tight and tense as an overwound clock, but I think perhaps I can begin to taste the underlying, unmistakably nebbiolo fruit beneath all the kicking and screaming. I begin to detect the faint but undeniable hints of a Barolo-to-be in a soiled swaddling cloth. I also notice, as I hand the glass back a second time, that the other fly is missing too!

From here, the contents are pumped out of the tank and into a pneumatic press with a long rotating cylinder. The liquid flows freely through the mesh filters of the press to holding tanks downstairs. Once much of the liquid has passed through, the remaining solids on the bottom of the fermenting tank enter the pump and begin to collect inside the cylinder of the press. Periodically, a big rubber cushion inside the core of the cylinder expands gently but firmly, pressing the solids to remove all the remaining juice. (This is called "press wine," or *vino di torchio*, and may be used to top off barrels of aging wine.)

Afterward, the dark purple, nearly dry solids (called *vinacce* in Italian, pomace in English) are shoveled out from an opening in the bottom of the press into huge square plastic tubs and later sent off to a distillery to be made into grappa (some people just dump them into a vineyard as fertilizer). As the fermentation tank empties, we rotate the paddle a few times to scrape the grapey crud off the sides and down to where it collects with the rest of the sedimentary sludge in a trough along the bottom. This trough has a screw conveyor down its entire length, which channels sludge out the opening, and it too is pumped to the press.

When the fermentation tank is empty, it is thoroughly hosed down with hot water. Once the last grapes of this season have been turned into wine, the real work of these machines is finished until next year.

After passing through the press, the new wine stays in tall vertical holding tanks in the basement for five to six days to rest and recuperate. During this time, it is periodically pumped over to aerate, allowing the new wine to "breathe" and residual gas to escape, and moved from one tank to the next in order to remove the thick sediment of particles and dead yeast cells, which collects on the bottom of the empty tank. (This sediment is scraped into plastic tubs, where, eventually, the liquid and solids will separate. The liquid is siphoned off into large glass demijohns and like the *vino di torchio* may be used to top off barrels during aging.) The practice of removing accumulating sediment is called racking.

After this week-long period of rest and racking, the wine is moved to wood casks for maturation. In these casks, the wine undergoes a second fermentation called malolactic in which harsh malic acid is transformed into softer lactic acid. This is a less radical fermentation than the primary alcoholic one but an important step in the wine's maturation process. Aside from lowering the levels of acid and alcohol, it helps enhance the softness, roundness, and complexity of a wine. Usually, the malolactic fermentation will occur spontaneously, though it too can be induced if necessary. Young, simple white wines such as a favorita or pinot grigio, in which freshness, liveliness, and pronounced acidity are desired, will generally be vinified in stainless steel at a low temperature to prevent the malolactic fermentation from taking place.

Then the "big sleep," a sort of gestation period, begins. What exactly is resting in these wooden vessels? Is it wine or merely some liquid-in-limbo on its way to becoming wine? From the standpoint of the transformation of fruit juice into alcohol, it is unquestionably wine and fermentation the moment of its birth. But from the standpoint of wine itself, fermentation is merely the moment of conception, the time when Yeast got it on with Sugar, took a tumble in the tank, and planted the seeds of Alcohol. At this point, the fermented juice is more like a growing fetus than a baby. It is not actually aging yet; it is still developing and maturing, still waiting to be officially "born" and brought out into the world. Technically it is wine—the essential chemical transformations have taken place—but it is not yet a wine you would enjoy drinking, and it is certainly not Barolo. The DOC laws for Barolo stipulate that following primary fermentation the wine must spend at least two years in wood casks—any kind of wood cask, from the large traditional *botti* of Slavonian oak to the small 150- or 225-liter French *barriques*—and after that, an additional year of aging wherever you want, either in bottle, stainless-steel vat, or oak cask.[3] (A Barolo "Riserva" must age for five years.) Until these criteria have been met, it may only be referred to as Nebbiolo *for* Barolo.

In the large cellar two levels beneath the apex of the Bussia, the barrels full of new wine are lined up in straight rows. They sit there in the cool, quiet, softly lit cellar like newborns in a nursery or larvae lying in cocoons. During this time they are constantly monitored and cared for: while the malolactic is taking place, we walk up and down the rows loosening the plastic stoppers in the barrels to let gas escape; each time, a hiss comes rushing out like a sigh of contentment. Sometimes we don't get there in time; the cork is violently shot out of the cask like a baby's sudden scream for attention and we need to find it, wipe off the splattered wine around the mouth of the barrel, and stick it back in. With wines that are being aged on their lees,[4] we insert a long metal rod with a curved tip on one end and a handle on the other and gently scrape the bottom of the barrel to stir up the fine residual yeast cells that are still being slowly consumed by the maturing wine, a technique known by its French name, *bâtonnage*.

The trauma ward of harvest and the chaotic birthing center of fermentation have yielded to the calm of an ordered nursery. Once the wine is in the barrel, tasks become much more routine and mundane. Wines in stainless-steel tanks are periodically pumped from one to the next to aerate and remove sediment, wine in barrel is either racked or stirred up with the metal rod, and remaining barrels still undergoing malolactic fermentation need to have the plastic corks removed to let gas escape. Used barrels are washed out, wooden pallets are stacked up outside, and rinsed crates are stored away.

Parusso, like most producers, makes a number of different wines, and each of them requires a different period of aging. This affords me the unique opportunity to taste different vintages of the same wine (Barolo Bussia) at various points in its evolution.

In one corner of the cellar is the total production of Barolo from the dismal vintage of 2002. There are about forty barrels in all, a fraction of what there would normally be. And though they were born (pruned, harvested, vinified, and aged) to become Barolo, all but ten of them are going to be declassified, blended in with the regular nebbiolo wine, and bottled as Nebbiolo Langhe, a lesser designation. "I'd rather make a good nebbiolo than a shitty Barolo," says Marco, "especially following on the heels of the great vintages of the past seven years."[5]

With the unpleasant memory of the tank sample of the 2003-to be still clear in my mind, I taste this 2002, which has been blended with the other nebbiolo 2002 prior to bottling. It is good, very good. The fruit is bright red cherry, ripe and round but with lively acidity. The tannins are soft and supple and the finish quite clean. It does not have the structure, the elegance, or the complexity that you would expect from a Barolo, even one in its infancy. And you can sense that the simple, pleasant fruit qualities it does have might have been spoiled by yet another year in wood. But it is good wine and full of distinctive nebbiolo character.

Next I try the Barolo Bussia 2001, which has been maturing in small *barriques* for the past two years and is just about to be transferred to bottles, where it will age for another year prior to being released. A "typical" vintage in the Langhe 2001 had a long growing season, hot days alternating with cool nights, and moderate rainfall (no significant hail). The tannins are very pronounced, giving the wine a harsh astringency right up front as soon as it hits your palate. But it is clear that beneath this tough exterior lurks a real beauty. There is gorgeous dark fruit—blackberry, dark cherry, and currant—a hint of cocoa and a touch of licorice, but all are restrained, moderated, and integrated into the firm tannic structure, giving the wine intimations of aristocracy and stubborn elegance. It is still tightly wound; it seems practically bursting at the seams and bouncing off in all directions. But there are already murmurs among Barolisti that 2001 will turn out to be an exceptional vintage. This one will do quite well with the year-long repose in bottle that lies ahead.

The 2000, on the other hand, has been resting in bottles for the past year and a half; this spring, the bottles will be labeled and released to the market. And none too soon: the *Wine Spectator* has declared this the VINTAGE OF THE CENTURY, awarding it an unheard-of 100 points, and the market is once again clamoring for supplies.

The 2000 is good wine to be sure: fat and fleshy, almost sweet, with loads of ripe luscious fruit and soft, supple, almost creamy, tannins. This indeed is a Barolo for the masses: readily accessible, easy to love, and impossible not to like, it holds nothing back; it is unquestionably attractive and enjoyable if, ultimately, forward and perhaps a bit superficial. By contrast, 1999 is similiar to the 2001 but, with a bit less fruit and more pronounced tannic grip, it comes across as even leaner and more muscular. (I personally prefer the more austere structure and restraint of '99 or '01 to the highly touted 2000, but that, of course, is a matter of taste.)

This rare opportunity to taste the same wine from different vintages gives me a glimmer of insight into that first difficult encounter with what will become Barolo Bussia 2003. Of course, to really make any sense of it you have to do this over and over and over again, make tentative guesses and evaluations (and, in the case of a winemaker, critical choices), monitor the development before and after bottling and then, years later, see how it all panned out. Nevertheless, I push myself to hazard an assessment: 2003 Barolos will be big, forward, fleshy wines packed with ripe fruit and high in alcohol. Tannins, the musculature of a wine, will be relatively understated and hardly able to contain all the fruit so the wine may come off a bit unfocused and lacking in complexity. They should be perfectly lovely to drink young, but will probably start to decline after about eight years, just around the time the 2001s will be reaching their prime.

This, of course, is only a guess. But whatever happens, it will be interesting to see what becomes of the 2003 Barolo I tasted right out of the fermenting tank and remember what that very first taste was like.

Once the fermentation is over and the new wines are all resting comfortably wherever they need to be, things quiet down considerably in the cantina and activity sinks into an increasingly predictable routine. The primary tasks may be finished but there are a whole host of other chores to do—bottling wine, labeling bottles, assembling cartons, filling orders, cleaning, cleaning, cleaning. There's never a dull moment in the cantina. Well, that's not exactly true: there are plenty of them.

I am what is known in the industry as a "cellar rat." I spend hours going up and down rows of barrels letting out the last of the gas or stirring up the sediment on the bottom; long days outside in the cold rain washing out used *barriques* or inside helping move wine from one vat to another, scraping the leftover sludge out of the bottom of the tank, and then hosing it down.

The work is not always easy or interesting or fun, especially compared to the sunny elation of harvest out in the vineyards. In fact, it often borders on drudgery. Sometimes I feel like I'm being punished. From having my own business, fielding compliments from grateful customers, calling the shots, and being the boss of many, I find myself on the bottom rung of the ladder doing the most menial of chores.

For the most part I don't mind. I like the atmosphere of the cantina—the sweet, musty smell of fermented grapes, the pregnant hush of wine developing inside wooden casks, the Italian pop music blasting out of the radio. But all the same, I get occasional pangs of doubt: "What am I doing here, busting my butt hour after hour, day after day?"

"Where," I ask myself with increasing urgency, "is all this going?"

· RECIPE #22 ·
C U G N Á

Grape Must Conserve

CUGNÁ, A GRAPE must–based condiment made throughout the wine-producing region of Piemonte, is an essential accompaniment to *Bollito misto* and aged cheeses. Nonna Rita's favorite way to eat it is with polenta. Because it is sweet but not too, it is extremely versatile and straddles the fence between sweet and savory: it can be served with full-flavored game for dinner and again for breakfast the following morning with toast or croissants.

As the must is literally snatched away from the fermenting vat just prior to vinification, Cugná represents an alternate journey or transformation of the fruit otherwise destined to become wine.

Following is Nonna Rita's recipe. Actually, it's more of a procedure than a recipe and an easy one at that: Rita has nothing written down, and it changes a bit each time she makes it. Ingredients vary greatly, and everyone has their own preferences. For these reasons (plus the fact that Rita uses 50 liters of must at a time), the quantities and ingredients given are mere suggestions.

INGREDIENTS

3 gallons grape must (see note
below)
1 pound of small firm pears such
as Seckel, peeled, cored, and
cut into pieces (see note
below)
2 quince, peeled, cored, and cut
into pieces

½ pound (8 to 10 pieces) fresh
figs (preferably white),
whole, stems removed
1 cup shelled hazelnuts or
walnuts, lightly toasted
(see note below)
1 small lemon, quartered, seeds
removed (optional)

1. Place the grape must in a large, heavy-gauge pot and bring to the boil.

2. Lower heat and simmer and cook until it begins to thicken and is reduced by about two-thirds. This will probably take three to four hours. (Rita says that the "better" the grapes—that is, the more natural sugar they have—the less time it takes to reduce.)

3. Add the fruit and nuts and lemon. Allow to come back to a simmer and cook until the liquid is reduced by about half again. You should have about three quarts. The Cugná should be dense and syrupy, the fruit slightly broken down though still somewhat chunky like a thick preserve. To test for doneness, spoon the hot, dense liquid onto a plate; if it runs, cook a bit longer.

4. Ladle the hot Cugná into sterilized glass jars with a rubber gasket lid and close securely *(see note below)*. Let cool gradually, then store in a dark cool place until ready to use.

YIELD: MAKES ABOUT 10 TO 12 PINT-SIZED JARS.

———————

NOTES: Grape must is unfermented grape juice obtained from crushed wine grapes prior to vinification. It is available during harvest time from wineries in any wine-producing area (of which there are more and more throughout the United States). Because the must should be fresh, unpasteurized, and unadulterated, this is a recipe that can only be made in the fall. Once made, however, it lasts a long time (see below about nuts and storage). For this reason, along with the extended preparation time, it makes sense to make a large batch of Cugná to enjoy through-out the year. Use whatever kind of must you can get your hands on—any kind will do, though some work better than others. Rita Parusso has made Cugná from all the different grapes they have: her favorite is dolcetto (lots of bright, forward fruit; soft tannins). Nebbiolo works well too. She cautions against barbera because of its high acidity and says that even white grapes such as sauvignon blanc or chardonnay will work, although the must will turn dark because of the long cooking.

For pears, Rita uses Madernassa or Martin Sec (so called because they are traditionally harvested on Saint Martin's Day, November 11, which also happens to be the day when farmers were traditionally paid). Both are small, hard, cooking pears that hold up, and in fact release their flavor best, over long cooking. Both of these varieties are late-ripening, coming into fruition right around harvesttime. Use Seckel pears or any other small, firm (but ripe), tasty variety you can find.

While nuts are a standard ingredient in most Cugná recipes, Rita cautions that the nuts may go rancid (without them the Cugná will last indefinitely). When she does use them, she uses the indigenous Tonda Gentile hazelnut, which is smaller than most other varieties and has a more pronounced flavor.

3 I · SUBTERRANEAN TREASURES

AFTER THE BUSTLE of spring in the vineyards and the frenzy of har-vest, a quiet calm descends over the Langhe. The long hot days of summer seem but a distant daydream. The new wine rests comfortably in barrel or vat while outside the days grow ever shorter and cooler. The pressure is off; there is time now for other activities. And one of them is searching for truffles.

There's a lot of mystique surrounding this homely-looking tuber.

To begin with, truffles are truly wild and, despite many attempts to cultivate them, they have stubbornly remained outside the bounds of control. Truffles are ridiculously expensive (they are commonly considered to be an aphrodisiac), and are available fresh for only a few months a year. There are many products that try to hijack the truffle's allure—oil, cheese, butter, and pasta—but contain mostly chemical flavorings.

All this fuss over a bumpy beige-brown fungus.

But the flavor! There is simply nothing quite like it. Ethereal and earthy, pungent and powerful yet delicate, nutty with a hint of ammonia, there is nothing to compare with the flavor of the fresh white truffle. Actually, the appeal is as much about the smell as the flavor, just as it is with a great wine. But these unique qualities are elusive and can be easily scared away. For this reason, Piemontese are purists when it comes to truffles. They are never cooked but shaved raw tableside with a special handheld shaver. There is a visual stimulation in this, and an aromatic one too. As the transparent shavings alight on the food, the pungent perfume of the truffles is released and the two mingle merrily before your eyes and in your nostrils prior to dancing on your palate.

There are certain dishes that are considered to show truffles off best: fresh tajarin noodles tossed with butter (no sage in this version and don't wait for cheese, because it is usually not served with truffles); raw meat, either chopped or thinly sliced and dressed with a delicate Ligurian olive oil and salt; a sunny-side up egg *(occhio di bue);* a simple risotto or Fonduta. Most any other paring is considered wantonly risqué, potentially wasteful, and philistine.

Make no mistake, we're talking here about the fresh Alba white truffle. Black truffles come mostly from Périgord, France, as well as from Umbria in Italy. They are found around here too but are looked down upon as inferior relations, the black sheep of the truffle family. Black truffles are less aromatic than white ones; they are less rare and therefore less precious, costing about a third what the white truffles do. Their thick, gritty skin is usually peeled (the peelings are good to infuse flavor into a sauce) and they are best when cooked; rather than spoiling it, cooking releases their flavor.

The French use pigs to find truffles, whereas here in Piemonte they use specially trained dogs. Being a cook, I had long heard the lore of truffle hunting and had a preconceived idea of what it would be like, just as I had with picking grapes. Once again my expectations turned out to be pretty inaccurate. Truffle hunters tend to be rather secretive, solitary characters, and they like to keep a low profile. But I have a friend who is a *trifulau* (as they are called around here), and he agreed to take me along on one of his many autumn expeditions.

Fiore, Truffle Hunter

FLAVIO (HIS REAL name is Claudio but everyone knows him as Flavio) lives in the Bussia with his sister, my friend Bruna, and his younger brother, Aldo, and his mother, whose real name is Elena but whom everyone knows as Maria. None of the children ever married. Flavio is the man of the house, the uncontested holder of domestic authority, inherited from his father along with his nickname, *Fiore,* which means "flower." Aldo is known as *Fiorino* (little flower).

Fiore does different things at different times of the year. Everything has its time. During tourist seasons he dons a crisp white shirt and tie and serves food in the fancy restaurants of Castiglione and Monforte. In the spring he tends the vegetable garden and family vineyards. Throughout the year he sells insurance and is always ready to call on a prospective customer, day or night. And in the autumn he is a *trifulau.*

As a child, Fiore went truffle hunting with his father, who taught him how to talk to the dogs and, more important perhaps, to listen to what they said. He has been doing it ever since. (Aldo takes care of the commercial side.) Call it what you want—hobby, pastime, seasonal occupation—truffle hunting, while never certain, can be a lucrative activity, though if you divide the total revenue generated by truffles by the hours it takes to find them, it might appear less so.

Trifulau, however, don't calculate their income by the hour. It's not a job to them. It is just something they do, an irresistible seasonal compulsion. Truffles are there for the taking, there all around just below the surface, near the roots of trees in the woods between the vineyards, in the crevices of the hills, yet another bounty the earth offers up freely, seductively. All you have to do is find them.

I was a bit surprised at first when Fiore told me to come by late one afternoon. When I get there he is sound asleep on the couch in the living room; apparently, he has been out all night in the woods. Maria gives him a rough shake to wake him up. He jumps up, surprised and a little disoriented, smooths back his thinning hair, and tucks his wrinkled shirt in under his bulging belly.

"Hey," I say, "we can do this some other time, you know," as I follow him out the door.

"No no, va bene, andiamo," he replies unenthusiastically, climbing bleary-eyed into a dirty gray four-wheel-drive. I hop in on the other side, the motor grinds into gear, and we set out.

"Anyway, don't you have to go for them at night? I thought truffle hunters prefer to go at night when no one can see their secret spots."

"No," he replies, gradually gaining coherence as we crawl up out of the Bussia. "You know, the truffles are there during the day as well as at night. And where they are is no secret. They're in the woods over there, or over there," he says, pointing to the right and left. "You just have to *find* them. There's no magic. Everyone knows there are fish in the sea, right? Just because you caught one somewhere once, doesn't mean there'll be more there the next time you go back."

He continues, "Now, of course, there are some spots you're more likely to find them, out-of-the-way places where there are more oak or chestnut trees, where the ground has been less disturbed or trampled (the woods are disappearing, you know, more and more every year). But it's no secret really. Look, see that car over there at the edge of those trees? That's a *trifulau*. And over there's another one. When you see a car parked by the edge of the woods, it's either a truffle hunter or a couple *imboscata*." He chuckles. (Unless the vehicle is old, beat-up, and mud-splattered, it's more likely to be a pair of lovers.)

"They all know me and I know them; we say *'salve'* when we cross paths but, just like fishing, you don't want to be right on top of each other. We like to leave each other alone, give each other room to breathe; professional courtesy, you know. But it's no *secret*."

Fiore is opening up now and warming to the subject. He goes on without prodding. "There are some advantages to going at night. It's quieter; there are fewer distractions for the dogs. They can concentrate better. Some people say there are fewer competing smells so the dogs can pick up the truffle scent better, but I think the aroma actually comes out more in the warmth of day than in the cool of night."

"Speaking of dogs," I say, "don't we need to bring one along?"

"Yeah, of course. We did. Mickey's there in back."

I turn and look over the backseat into the cab. There, curled up and silent, is a black muscular mutt, one of a multitude of dogs usually tied up outside around the Alessandria compound. Mickey is Fiore's best truffle dog. He has been trained since he was a pup and showed great promise from early on—an incredibly sensitive nose and an incessant love of the hunt. A good truffle dog is extremely valuable; jealous *trifulau* have even been known to steal or poison a talented competitor. But it is not just a matter of training; some dogs have it and some don't. And Mickey definitely *has* it. He knows where he is going and what he is about to do and is immersed in a deep power nap like a prizefighter before a bout.

We pull off the asphalt road above the Bussia Soprana and chug along a dirt trail to the edge of a forest. "Let's try here," says Fiore, pulling the battered vehicle off to the side of the path.

Fiore, with well-worn boots, short-brimmed wool cap, and a multi-pocketed vest over plaid flannel shirt, grabs his round-handled wooden cane with metal tip—the essential accessory of every truffle hunter—and lowers himself down out of the dirty Jeep. We go around to the back and Fiore opens up the hatch. Mickey comes bounding out and makes a break for the woods. The previously catatonic animal is off and running.

Flavio and I amble up the narrow dirt trail into the woods and soon catch up with Mickey, who has been making frequent forays into the dense forest on either side. Then the real hunt begins. I had imagined truffle hunting as little more than a pleasant walk in the autumn woods, but it is actually work, a shared, concentrated endeavor of man and dog, and Fiore and Mickey go at it in perfect symbiosis.

"*Dai, su! Forsa, forsa! Su!*" encourages Fiore. Mickey responds with an even greater burst of energy. Thus begins a constant and obviously familiar dialogue, the words interspersed with clicking sounds and whistling. This is their private language, one reserved solely for this special activity.

"*Baica bin, Mickey. Baica, baica bin!*" Naturally, Mickey speaks Piemontese.

Up and down, forward and back, right and left he goes, darting here and there, nose to the ground, sniffing feverishly, egged on by his master. The dog flits around nervously and continuously like someone who just can't get comfortable, can't find the right spot. Then something catches his interest. He zeroes in and starts to paw the ground.

Fiore lets him dig for a moment or two before giving a firm command. *"Ven si, Mickey, VEN SI!"* His master's no-nonsense tone pulls him away from the spot and back to the trail, where he receives a treat from his coach, a piece of stale bread. Mickey immediately makes a beeline back to the shallow hole and resumes digging with renewed vigor.

"Looks good," whispers Fiore. "When he goes right back to the same spot, it usually means there's something there."

" 'Spetta, Mickey. 'Spetta! Vai via!" shouts Fiore, approaching the spot and tossing some pieces of bread off to the side. Mickey retreats, collecting the treats as the *trifulau* lowers himself to his knees and begins digging with his cane. Periodically he stops to sniff the tip, then takes a handful of earth and sniffs that too.

"Smell that?" he asks, holding a handful of damp, cool dirt up to my face.

"Yes, I sure do!" I say. The smell is unmistakable; the earth is perfumed with the precious, intoxicating aroma of truffles.

Fiore takes a small trowel out of his pocket and begins digging, periodically lifting handfuls of dirt to his nose. He soon puts the trowel down and continues scraping away at the earth with his thick fingers, tossing the occasional piece of bread off to the side to keep the irrepressibly inquisitive Mickey at a distance.

Then he stops.

"Ecco. You see?" he asks, sitting back on his knees. I look over into the hole and there it is, about six inches below the surface, a brown nugget nearly the size of a golf ball, cradled in but clearly distinct from the brown earth around it; a truffle resting in its damp subterranean womb.

Fiore pulls out a little brush and, like an archaeologist excavating a precious relic, brushes the dirt off the sides and bottom of the dusty object to reveal its true dimensions; he doesn't want to risk yanking it

out and breaking off a valuable piece. When the truffle is fully exposed, he lifts it out gingerly—it doesn't apear to be attached to
anything—and does a careful examination, smelling it and gently
squeezing it with his thumb and index finger.

"Hmm. Nice and firm. Fresh too. This one probably formed in the
last few days, maybe right after that rain we had on Sunday. And it's a
pretty good size, no? *Bravo, Mickey, bravo,"* Fiore praises profusely, holding his fist with the truffle in it close to the dog, allowing him to deeply
pull the aroma into his dirt-crusted snout. Fiore shovels the dirt back
into the hole, strewing leaves over the top to cover his tracks and,
standing up with the help of his cane, wraps the prize in a small piece
of newspaper and places it in a big side pocket of his vest. "Bravo,
bravo, Mickey!" The dog too stands up on his muscular hind legs, eagerly soaking up his master's pats and praises along with more well-
earned treats of stale bread.

With that, Mickey's off again and so are we. *"Baica bin, baica bin,"*
urges Flavio. *"Dai, forsa, su. Su!"* The scene repeats itself over and over
again as Fiore's pockets begin to bulge with lumpy tubers.

There are some misses: Sometimes, the prize turns out to be no
more than a minuscule pea-sized pellet. Too small to slice, almost impossible even to clean, these go to the dog as a special treat, a special incentive to find more, and the hunt resumes with renewed enthusiasm.
And there are some occasions when, despite Mickey's pointed interest
and Fiore's careful excavations, nothing at all is found. "There *was*
something here," insists the disappointed *trifulau,* standing up and
dusting off his knees. "It probably just rotted away."

Most often, however, when Mickey shows interest in a spot, a truffle
is found there. We hit three different areas, spending a total of about
four hours in the woods. The take is not bad; seven truffles, total. Four
of them are salable, while three are too small and knobby to be worth
much at the market. The first one we found turned out to be the
largest of the day's catch; the small ones Fiore gives to me.

Back at the Alessandria homestead, we enjoy a *merenda* (afternoon
snack) of bread, cheese, and rough fizzy homemade wine while
Fiore/Flavio/Claudio weighs each truffle and adds them to a shoe box
piled high with recent quarry waiting to be sold.

*   *   *

Each October, there is a big truffle fair in Alba, the self-proclaimed white truffle capital of the world. Banners are draped across the main streets, bottles of wine grace shop windows, and the stores stay open late to cater to the throngs of tourists who pack the hotels, restaurants, and cafés. At the center of it all, under a big tent in a large *cortile* right off the Via Maestra, is a special market where truffles are available for purchase by the general public.

The irresistible allure of the white truffle attracts people from all over the world, and this means big business for many, many people besides the humble *trifulau* who find them.

The only problem is that this year there are no truffles, or at least very few. The intense heat and drought of the past season have created conditions decisively unfavorable for this golden fungus, which loves the cool and damp. A good year for wine, goes the common wisdom, is a bad one for truffles (and vice versa). Truffles thrive in the autumn following a cool, damp, wet spring and summer, which this one definitely was not. Truffle hunters are finding very few of them this year and those they do find are smaller, drier (with a tendency to crumble), and not quite as pungent as they should be; prices, in reverse proportion to the quality, are exorbitant.

The fair, however, must go on. I walk through the rows of vendors in the market one evening with Fiore, who shakes his head in disgust. Two-thirds of what he sees there are impostors, truffles smuggled in from Albania or Czechoslovakia or even China masquerading as Alba white truffles. They are not as aromatic, the flavor is not as intense, the contour generally smoother than the real thing. It is hard for me to tell the difference just by looking at them but Flavio can pick them out a mile away—"This is not ours, nor this, nor this. This one *is* ours; see how it has that bumpy exterior and dark color? This one no . . ."

"As far as I'm concerned, they can bring in truffles from wherever they want," Fiore gripes on the way out. "Just don't try to pass them off as ours."[1]

"Maybe they don't know," I suggest.

"Oh they know, all right. Everybody knows, from the people behind the booths dressed up like *trifulau* (they're not really hunters, they just pretend to be; really they are just merchants) to the big shots who organize the fair. There's just too much money riding on it. No one who knows truffles buys them at the official festival market anyway."

"Oh really? Where *do* they get them then?"

A few days later, I go along with Aldo to unload the contents of the shoe box. He has an official truffle seller's license and a business card with a color picture of himself on it sniffing a huge white truffle. We drive through the chilly darkness of a predawn morning like smugglers going across a border to the little town of Ceva, about forty minutes away. We park the car and walk to a nondescript piazza where vendors are setting up their stalls for the weekly market. But we don't stop.

"We're going over there," says Aldo, indicating an even more non-descript parking lot off to the side of the piazza. "Uh, okay," I say, suppressing my confusion.

"I thought we were going to a truffle *market*," I say to myself, "not hang around a parking lot in the cold." Except for the nearby café, which is doing a brisk business in coffee, sandwiches, red wine, and the occasional grappa, there doesn't seem to be much of anything going on here, just a bunch of men standing around with their hands tucked into their down jackets and shoulders hunched up against the morning chill.

Then something happens: two guys dart around to the back of a car, where one of them pulls something out of his pocket, though I can't see what it is because a tight circle of people immediately forms around them. After a moment, the group disperses. Then, about ten minutes later, it happens again right near us. This time I see a guy take a brown paper wad out of his pocket and unwrap it, holding the precious contents up for inspection. The intense aroma of fresh truffles, lots of them, wafts through the brisk early morning air, obliterating the smell of coffee from the nearby café. The deal is transacted quickly; the second guy stuffs the brown wad into his coat while the first tucks the rumpled paper bills into his pants pocket and the group adjourns.

The same thing happens yet again a few minutes later and then again and again.

Aldo holds back, monitoring the transactions with the intensity of a seasoned Wall Street trader. He is convinced prices will rise as the morning progresses and the slimness of the pickings becomes apparent to all. It is hard for an outsider to distinguish the buyers from the sellers until a deal actually goes down, but they, of course, all know each other and who is doing what. The people with truffles to sell are mostly the people who found them—that is, farmers and part-time hunters from the surrounding area—while those doing the buying are primarily local restaurateurs, shop owners, or wholesale brokers. Milling through the crowd are also representatives for the handful of large companies that make truffle-based products and export truffles to the United States.

Around ten, the pace begins to quicken and Aldo jumps into the fray, emptying the contents of the shoe box in under an hour. Prices overall are good, between 250 to 300 euros per 100 grams. And this is a tax-free transaction for both buyer and seller. In reality, the white truffle market is mostly a gray one; unlike Aldo, few of the casual hunters and small-scale dealers have the official license to buy and sell truffles, and most do not declare the income from them. *That's* why people tend to be secretive. The authorities surely know what's going on and usually let it slide, but no one wants to tempt fate by being unnecessarily brazen.

Just before noon the market evaporates as suddenly as it started and we head for home, Aldo clutching the empty but still fragrant shoe box, his coat pockets now bulging with cash as his brother's had with truffles. My pocket bulges too with a rumpled package of small but firm and fragrant truffles I managed to pick up for a song of eighty euros, and plan to use (generously!) for a private first-anniversary truffle dinner at home the following week.

32 · DISTILLATION

BACK AT THE cantina, with the fermentation, maceration, and pressing of the wine now long over, it's time to shovel all of the accumulated *vinacce* from the large square storage containers into plastic drums and send them off to the distillery. And this reminds me it's time to pay a visit to my old friend Romano Levi.

Levi's distillery is located just outside the town of Neive over in the Barbaresco area. There is no telephone there, so you can't call to arrange an appointment, and no sign outside. Yet somehow, even the very first time I went there, without knowing exactly what I was looking for, I knew when I had found it.

Distilleria Levi Serafina

I PUSH A nameless plastic doorbell outside a wide iron gate and hesitate, asking myself whether I am really in the right place. Several minutes later a little bald guy comes swiftly out from around the back, across the threshold and up to the other side of the gate. He opens a door in the gate, but only a crack. He has incredibly thick, long eyebrows that shoot straight up, framing small wizened eyes. "*Sì?*" he inquires. "Uh, I'm here to pay a visit," I reply. He wipes his chin for a moment, then steps back to let me in, closing the door behind me, and walks swiftly back across the courtyard, saying, "*Vieni, vieni, vieni.*"

At first, I assume this man is merely some pensioned porter. But once inside, I realize by his proprietary air that this is none other than the legendary Levi himself.

I follow behind him, looking with barely concealed amazement and curiosity at the strange place around me. It is more of a compound than a factory. In the rear behind a wildly overgrown garden is a nondescript, single-level house where Romano lives with his elderly sister. On the left is an old brick building housing the offices, and on the right, the anonymous wall of an adjacent warehouse. In between, in a haphazardly formed courtyard, is a small rivulet flowing into a drain, and next to that is an assembly of thin, vertical rocks arranged in a gridlike pattern with a pair of tall skinny ones towering above the rest.

"Hey, that looks like New York City!" I exclaim.

"Oh, you're right, indeed. That's just what it is, the *Grande Mela* [Big Apple] right here in the Langhe. See? There are the *Torri Gemelle*, the Twin Towers. Ha! New York on the Tanaro!"

The place is like an improvised art park with little sculptures all over the place, poems tacked up on the walls, and miniature paintings on terra-cotta roof tiles. And there are owls, owls everywhere, in a multitude of different sizes and materials. It suddenly occurs to me that Romano's tiny eyes framed by big bushy brows are, in fact, those of an owl.

I try to keep up as Levi zips quickly around the corner of the old brick building, past a tall open-roofed wooden structure under which blocks of dried-out stuff resembling peat moss are stacked to the rafters, and into the distillery.

It's extremely warm in the small room, and the distinctively heady smell of *vinacce*, a sweet-sour blend of fresh earthy manure and turpentine, is almost overwhelming. Against the far wall, actually built into the wall itself, is the old copper still. On its face is a circular portal with a rotary latch, and in front of that a wooden press over a drain; the cement wall nearby is splashed with dried red. Just behind the big copper kettle in another room is the brick oven that heats it.

A tall ladder leans up against the back of the still, and someone is on top tinkering with something. A copper tube comes out of the domed roof of the still and goes spiraling across the room, branching off to three raised wooden barrels over to one side. On the other side of the room is another barrel with a wooden lid resting on top. The walls are covered with moldy old photos, letters, and more owls.

"So," Romano says, "would you like to try some grappa?" But he doesn't wait for a reply (I can't imagine anyone coming here would ever say no). He lifts the lid and lowers a little glass beaker tied onto a long string into the barrel. The glass goes "Ploop!" when it hits the grappa, and "Ploop!" mimics Romano gaily as he pulls it back up, like a merry fisherman reeling in his catch.

He offers me the little cylinder, and I tip its contents back into my mouth.

The first impact is a burning that soon mellows into a sensation of spreading warmth as it travels down the throat into the chest and stomach and, at the same time, up through the sinuses into the head. There is a certain coarseness, to be sure, but there is smoothness and balance as well. There is also an undeniable assertiveness and genuineness: this *is* what it is, and what it is is *not* something to be taken lightly. This is *real* grappa.

There are other wonderful distilleries—Berta and Marolo here in Piemonte, Jacopo Poli and Nonino in the Veneto—that make excellent grappas that are much more refined and aromatic and elegant than Romano's. Some come in beautiful, hand-blown bottles, and many are distilled from fresh grapes or a variety of other fruits rather than raunchy pomace. But this is something else, something humble yet noble at the same time—this is the essence of *vinacce*, the leftovers of fermentation.

Romano disappears (he never stands still very long), leaving me to contemplate this odd setup that resembles an out-of-control chemistry experiment in someone's garage.

The *vinacce*, brought here by winemakers throughout Barbaresco and Barolo, is put into the still, and water is added to rehydrate it. As the still is fired up, the mix inside heats up too, creating steam. In the closed container, the steam becomes condensation—an extremely pure and highly alcoholic "spirit"—which travels up and through the copper tube in the top and into the barrel.

Real grappa is a cleansing, concentration, reduction, and fortification; a distillation of the essence or spirit of what is left after the grape becomes wine. You can apply the process of distillation to other fruits, grains (wheat, rye), or vegetables (potato, as in the case of vodka), but you can make genuine grappa only from *vinacce*, the pomace of grapes that is left after fermentation.

In some ways distillation is similar to fermentation: it is, fundamentally, the transformation of one thing into another. But it is different too. It is a secondary process, a continuation of the long transformative journey from grape to wine, and grappa is the end of the line. Some might even say that distillation is a different trip altogether, an afterjourney, a sort of revival or wake, since most of the grapey characteristics got up and left at the previous stop.

Romano returns and offers me another taste of grappa from the barrel. "Ah, you like it? I am glad." He pauses for a moment, rubbing his hands together and staring down at the splattered floor of the distillery. Finally he makes a statement: "Grappa is the act of turning nothing into something; of making something good out of the leftover, that which would otherwise just get thrown away."

I ask him about the mountain of dried bales outside. "Oh, that. Yes, well, you see that press there? When the distillation of the grappa is finished, when there's nothing left to get out of those old grape pressings, we turn the wheel and open up the still. The remaining water flows through the press to the drain, and we squeeze all the liquid out of the solids to form those rectangles. We then stack them outside and let them dry out. And we use that dried-out leftover of a leftover to heat up the still; that's what we make the fire with, ha! And do you know what we do with the ashes? We give them back to the winemakers to fertilize their vines. It's a big circle, see?" he asks as he traces a circle in the air with his gnarly finger. "Ha! Well, how about a taste?" I don't say no.

The doorbell rings, and Romano scuttles out to respond. He lets some more people in and shows them around. At a certain point (I have been there for hours and have lost count of the number of times the beaker has been lowered into the barrel) it is time for me to go, and I say good-bye to Romano.

"So, is there something else we need to do?" he inquires, his hands clasped in front of him.

"Well, I wouldn't mind some grappa to take home with me, if that's possible."

"Ah, yes. Yes, I think I have a bottle or two I can let you have."

I follow this little wizard around the corner and into his office, which is like something out of a Dickens novel. It is dark inside, with late-afternoon sunlight shooting in a single smudgy window. Everywhere are stacks of old newspapers, piled high, yellowed with age and covered with a thick layer of dust. Big, dense cobwebs hang from the walls. Over in the corner near the window is a desk, and on the desk a cup full of pens, an assortment of different colored inks, and a small stack of hand-torn pieces of white linen paper. A shelf nearby is lined with thick clear glass bottles of grappa.

Romano sits down at the desk and places a pair of glasses low on his nose. "Now let's see, what shall I do for you?" (Romano creates whimsical hand-drawn labels for his bottles of grappa that are as prized as the grappa inside them.)

"Whatever you like," I answer.

He takes a piece of the white paper and begins to draw with the different colored pens in simple, clean, almost childlike lines: green mountains on one side and tall red buildings on the other with a blue river running between them. Above the mountains, a yellow sun is shining, and dancing over the river is the stick figure of a woman holding a bottle in one hand and a flower in the other. Along the top is written:

La Donna Selvatica delle Colline Fa Una Visita a New York sul Tanaro
(The Wild Woman of the Hills Goes to Visit New York on the Tanaro)

He wraps the bottle up in a piece of old newspaper and accepts 20,000 lire (about thirteen dollars) in exchange. I follow him back out into the sunlight, and he shows me out the gate—but not before passing back through the distillery for another dip into the barrel.

That was about fifteen years ago. Now, during the present winter lull, I go back for a visit with Bruna, who has, of course, heard of Levi but never been there.

Some things have changed since my first visit: Romano's reputation has, deservedly, grown and spread. This time we are ushered through the gate by an assistant. As we cross the courtyard, I notice the towers of stone are missing and ask what happened to "New York." "Oh that," he answers. "We had to take it down because kids couldn't help playing in it and would slip and get hurt."

When we get around back to the old still, there is a film crew from

RAI (Italian state television) shooting a segment on Levi. An exhibition of his famous labels along with mementoes and other paraphernalia from the distillery is going on in the nearby town, and colorful banners announcing it adorn the streets. Romano himself is as sprightly and impish as ever, but his hair has turned totally white and his long bushy eyebrows have been trimmed. He moves a little slower than he used to.

"He is well but tired," his assistant, Franco, says. "The show is a great tribute, but it has taken a lot out of him."

He can't draw all the labels by hand anymore, but we do manage to get him to draw a simple one on a bottle for Bruna. As he does, I notice that the office is now (relatively) organized and dust free. Apparently, the government agency that issues distilling licenses cracked down and made him clean it up. And back in the distillery the beaker on the string is gone. "Sanitary reasons" is the explanation; apparently you can't have people drinking out of the same glass and dipping it back into a barrel (even though any germs would surely be neutralized by the high level of alcohol).

Some things have changed, but some have not. Romano, though older, is the same. Miraculously, after all these years, he remembers me, the American with whom he posed for a photo in front of New York on the Tanaro. The old copper still with the red-stained wall in front and peat fire in back is the same. And the grappa that comes out of it is the same too (it now costs twenty euros). A bottle of it sits open on top of the closed barrel, and next to it is a stack of little plastic cups. "How about a taste?" Romano asks, handing us each a cup and pouring grappa into it without waiting for a response. He still doesn't need to: no one who comes here ever says no.

33 · TRIALS AND TRIBULATIONS

FROM THE WINDOWS at the back of Cantina Parusso, you can look down on to the handful of houses that make up the Bussia Sottana, and over to Fabrizio's section of Le Munie. The pale, bare vineyard stands in stark contrast to the dense dark green vines around it, reflecting the ever-diminishing daylight.

Italian is a gender-specific language. The common word for vineyard, *il vigneto,* is masculine, but in the case of Le Munie that seems inappropriate. Better is the term *la vigna* (feminine), which connotes a smaller vineyard or patch of vines.

This piece of earth is definitely female. And now it seems as if a green blanket has been yanked off to reveal a lovely white bit of flesh beneath, quivering in the fresh autumn air. Long-necked and high-shouldered, she reclines along the slope of the hill lounging coquettishly, un-self-consciously naked for all to see and admire. And admire they do.

"*È un bel pezzo,*" people say, barely concealing their desire. "That's a nice piece"—like a nice piece of ass. They would give it a proprietary smack if they could. Le Munie is coveted, sought after, desired, perhaps even more so now that she is stripped bare of vines and you can see her in all her glory: her voluptuous contours, her provocative position, her attractive reddish-gray complexion studded with stones and clumps of earth like freckles. Everybody wants her; to the extent that she is unattainable, they want her even more.

At a weekly market in the nearby town of Dogliani, I bump into a wine producer from Castiglione who knows I have been working in Le Munie. "Hey, when that Fabrizio gets tired of playing *contadino,* you tell him to come see me."

"I will indeed," I reply. "And if you ever get tired of working all that property you have, by all means let *us* know! We might be interested in taking on an additional piece or two ourselves."

But sometimes, stealing a glance from some tedious chore in the cantina, the naked vineyard takes on a different significance. It is no longer an object of desire but rather a barren, desolate wasteland; stripped down, vacant, blank, just sitting there doing nothing.

I am beginning to feel that way myself. Perhaps the novelty of my new life in Italy is starting to wear off, the reality of my present situation beginning to sink in.

I have an entry-level position performing basically unskilled labor for what in the States would be just above minimum wage. I go to work at seven in the morning when it is still dark and stumble home after seven in the evening when it is dark again, my long day filled with a succession of mostly mindless tasks like moving wine from one tank to the other and hosing out the dregs, washing out *barriques*, vacuuming, stapling cardboard boxes, or assembling wooden crates. Though the tasks may be simple, I often have to struggle to understand exactly what I need to do and struggle to make myself understood. The tables have turned. When I had my restaurant, I had many unskilled employees with limited English capabilities struggling to understand their jobs and taxing my patience; now I am the one taxing the patience of someone else. I have no medical insurance, I am without a car in a place where you really need one and without a steady job or a decent income, much less a career. I do have an apartment but it lacks the furnishings that would make it feel more like a home and I eat most meals alone on a borrowed chair.

Things were about to get worse.

Getting the apartment had been so easy; just sign the lease, get a bank account and a *codice fiscale* (like a social security number), and switch the existing utility accounts over to my name. It all went without a hitch. Then, about a month after moving in, I got a letter from the electric company. "*Egregio Signore,*" it began. "Welcome to our EGEA family. Please send us a copy of your residency in the town in which the premises are located within thirty days. Otherwise, we will cut off your electricity."

La Residenza. Every Italian has one. It is your base, your public anchor to the world, the place where your very existence is officially reg-
istered. After receiving the threatening letter, I went with Ivana to the
Municipio in Castiglione to find out how to get one. "Oh, it's easy,"
they said. "Just bring us your passport and your *permesso di soggiorno.*"

"Permission to sojourn" has such a nice, pastoral ring, like an afternoon lark in the country. In reality, it is anything but.

By law, anyone who does not have an Italian passport (or one from
another member of the European Community) must, within eight
days of arriving in Italy, go to the nearest *questura* (police headquarters)
and get a permit—*anyone,* whether you are planning to stay for two
weeks or two months. The permit doesn't entitle you to work or study
or anything else; just to *be* here.[1]

Of course, most visitors don't bother, and neither did I. I had already been here for extended periods without any problem and did not
want to tangle with Italian bureaucracy if I didn't absolutely have to.
Anyway, who ever heard of a permit to sojourn? But, faced with the
likely alternative of having no electricity in my new apartment, it appeared it was time to find out.

One afternoon, Ivana and I took a drive to the *questura* in the
provincial capital of Cuneo. We found a parking place right in front
and waltzed into the somber edifice, armed with my American passport, a bank statement (to show that I was not destitute), and a copy of
my New York apartment sublet (to show that I had some income and
so would not become a financial burden to the state).

Inside, a guard was sitting behind a beat-up old desk reading a pink
Gazzetta dello Sport. "*Buon giorno,*" said Ivana gaily. "We're here for a
permesso di soggiorno." Without looking up from his paper, the guard
replied, "Talk to him," indicating a harried-looking guy in rumpled
street clothes going in and out of an anonymous door to the side. On
his next pass-through, we managed to get his attention, waving our
documents in his face like a red flag. "*Scusi, signore,* we need a *permesso
di soggiorno.*" Irritated, he paused just long enough to say, "Yeah, well
you also need four photographs and a *marca da bollo* for ten euros
thirty-three. And hurry up; we're closing in an hour."

We huffed out, got the stuff, and were back at the *questura* in less than forty-five minutes. This time we used the doors marked UFFICIO STRANIERI, went right up to the front, and slid my passport, photos, and tax-stamp through the slot in a murky plastic window to an officer on the other side.

"Declaration letter?" he said, as he flipped through my passport.

"What?" asked Ivana, with obvious impatience.

"You need a letter from the person where he is staying offering to host him."

"What do you mean *where he is staying*?" she said, holding my lease up to the window. "He has his own apartment!"

"He CAN'T have his own apartment," the policeman said. "He's a *tourist*. Wait here."

"I can't believe this," Ivana groaned under her breath. *"Sono proprio cretini."*

Eventually, a door opened and we were ushered into the inner sanctum. Our "friend," the harried guy, descended a flight of stairs and came up to us. Ivana wasted no time:

"Ma per favore! First a *marca da bollo* and photographs, now some sort of declaration. I can't—" But his raised hand cut her off midsentence.

"Senta, signora. He has been in Italy for months without a *permesso.* Technically, he has no right to be here. I could hold him and deport him; they put a big red X on his passport and he can't come back for ten years. Do you want that? But if you leave right now," he said, looking at his watch (it was almost lunchtime), "I'll let you go. Get his passport stamped somewhere outside of the EU, and come back within eight days. And bring a declaration too."

We didn't press our luck any further and made a quick exit.

The following week we took a drive up to Switzerland. I managed to get my passport stamped at the border on the way back into Italy, but it wasn't easy, as they don't normally stamp anything anymore. Two days later, I headed back to the *questura* with my freshly stamped passport and a letter from Ivana declaring that she would host me while I was in Italy. (She wasn't off work again for a whole week, so I went alone.) I got there just after 6:00 A.M., as it was getting light. There were only a few people milling about outside. "This isn't going to be so bad," I thought to myself. "I should get my *permesso* and be home in time for lunch."

I went to a nearby café for an espresso, and when I got back ten minutes later, a crowd of people had formed: Africans, Moroccans, Latinos; young couples and older women with kids, but mostly men speaking a slew of Eastern European languages—Romanian, Macedonian, Albanian. I was clearly the only American around. On a ledge outside the courtyard was a wrinkled piece of paper with names and numbers written on it, but nothing else to say what it was. I went ahead and wrote my name down anyway. I was number 36.

As the minutes ticked by, the crowd grew larger, the courtyard filled, and tension built. Finally, at eight o'clock sharp, two guards appeared behind the glass doors and the crowd surged forward, pressing up against them on the outside. The policemen forced the doors outward, clearing a swath in the throng, sticking out their hands to keep people from spilling inside. People shouted and pushed, until one of the guards finally screamed "*SILENZIO!* We can either do this in a civilized way or we can not do it at all. We can just close these doors, and you can all go home. Would you like that?" An immediate hush fell over the crowd, and he continued:

"Okay, then. How shall we do this? Do you have a list or something?"

Someone handed the crumpled sheet to the guard, who took it without even looking.

"So. We follow the list? Is that what you want to do?"

"Yes, yes. The list, follow the list," said a handful of people with enough gumption to open their mouths. (I later learned that they almost always follow "the list" and that people had arrived as early as three A.M. to put their names down on it.)

The guard began calling off names from the list, five at a time, and people plowed their way through the crowd and inside the door. It was slow going. By eleven thirty, only the first fifteen names had been called. Once the last five had finished, the guard emerged and made an announcement: "Okay. Now we will take some *primi.* Anyone here for the very first permit, raise your hands." My hand went up, along with about half of the group. "Okay, I start to let you in, one at a time."

He might as well have said, "In a moment boiling oil will rain down upon you and the only safe place is inside these double doors," for the crowd lurched forward, many with their hands still raised in the air, pushing, shoving, and jostling for position.

"*Oh! Eh! Basta!*" said the guard. "*Basta. Basta! BASTA!*" The crowd grudgingly came to a stop.

"You're like animals, like children, eh? Okay, I treat you like children then. Get out of this doorway or I close it for good. Go on, get back. Make two lines one on each side, just like in school. You understand me? I know most of you don't; most of you don't speak Italian, probably never even went to school. But I'll give you a lesson right now. Move now, move over, or go home. Make two lines, single file, one over there and one over there."

He waited, standing solidly in front of the doors, arms crossed over his chest. "Come on! Do as I say. Pushing is not going to get you anywhere."

People reluctantly melted to either side of the door, but many continued to jostle for position (it was not true that pushing got them nowhere; it got them to the front of the line and in the door). I, however, retreated to a far corner of the courtyard. I was disgusted and humiliated. I was not going to fight some woman with children for a spot in line. I took a look around; these people were trying to immigrate to Italy to change their lives—perhaps to *have* a life—much as my grandfather did when he went to America. And here I was, reversing or retracing my grandfather's painful footsteps. These people wanted desperately to come to Italy in order to survive; I just wanted to hang out, to *sojourn* with Ivana among the vines.

"Let them go for it," I thought to myself as I kept to my corner. And I remained there all day. I was numb, exhausted, and dismayed. I didn't care about the *permesso* anymore—let them go ahead and turn off my electricity, let them deport me, I thought to myself. But I hung around anyway, mostly out of inertia.

Finally, at around five thirty I got up to and inside the double doors. Mustering whatever enthusiasm I could, I presented my recently stamped passport, my photos, my declaration letter, and tax stamp to the officer.

"Insurance?"

"Uh, what insurance?" I asked

"You need insurance. You have it? You can't get anything without it."

"Please," I said. "I've been here all day. I was here before. No one ever said anything about insurance."

"Well, what do you want me to do?"

"Um, could you just give me a *permesso di soggiorno*?"

No, he couldn't. I hung around inside the room for a while just to experience the minor meaningless victory of having finally, after ten hours, gotten inside. And then I left, dispirited, demoralized, exhausted, and empty-handed. It was one of the worst days of my adult life.

Several days later, while recounting my experience, somebody said: "You're going about this all wrong. *You* don't actually go to the *questura;* you need to send someone that 'knows' people there. I have a friend. . . ." The next day I met with Signor Umberto (the name is invented), who told me to send seventy-five euros to an address in Rome for insurance (he didn't know what it was, just that it was necessary) and bring him the receipt along with all my other documents. I gave him something for his trouble and had the *permesso* within a week. Actually, it was just a receipt; the real permit takes longer to print than the time it is good for. But that's all I needed. With that little slip of paper, I got my *residenza,* which entitled me both to electricity and to legally remain in Italy, at least for the next three months. But it was a hollow victory: the battle with Italian bureaucracy had left me bruised and battered.

Then something else happened.

One damp October evening shortly after getting my *permesso,* I emerged from a long day in the cantina into the fading light of dusk (*tramonto* in Italian) to see Fabrizio off in the distance walking through the bare ground of Le Munie. People were coming later in the week to plow over the vineyard, to smooth it out to blend in with the bowlike contour of the slope, and he was doing a preliminary walk-through. So I got in my borrowed car (Renata's son was out of town for a few months, and she had lent me her car while she used his) and drove over to say hi.

We chatted for a short time and then, as dusk fell, prepared to leave. My car was parked off to the side of the dead-end road a bit farther down, so I went first, backing into a bare patch and doing a three-point turnaround. Fabrizio was facing me in his car waiting to make the same maneuver. Perhaps for this reason, I made the turn just a bit wider than usual. (I had done this turnaround many times.) When I came to the second stop and prepared to go forward, I could feel that the rear of the car was on a slight backward slope. I cautiously switched into first, slid my foot from the brake to the gas pedal as I let out the clutch, and gunned the engine but did not move forward.

The little car had a pretty weak engine to begin with. And in that split-second when, as in many manual transmission automobiles, the car slid back slightly before lurching forward, the right rear wheel hit a patch of wet grass. Despite my foot on the gas, the car paused momentarily, as if considering what to do next, and then began to roll backward—slowly, gently, even gracefully, but uncontrollably. I could feel it happening but couldn't do anything about it. It was quite literally out of my control.

The car slid backward down into a gully, did a slow-motion flip, and landed upside down among the vines and posts of an adjacent vineyard. Once it came to a stop, I did a quick survey and found that I was, though shaken up and upside down, intact. Heart pounding, I released my seat belt and tried to open the door, but it wouldn't budge, so I crawled out through the open hatch in back. Fabrizio was there when I got out.

"Are you okay? What happened?"

"Yeah, I'm okay. But look at the car!"

I had never rolled over in a car before. I wasn't hurt, but I was stunned, pissed, and ashamed. I had done that turn so many times before; I didn't know how this had happened or why. Luckily there was no one else involved, I had not had even a drop of wine all day, and I did not need to go to a hospital; it wouldn't need to be reported to the police or to the insurance company. But that was hardly a consolation. I almost wished I had been injured instead of the car. It wasn't even mine.

Fabrizio and I drove over to Cantina Parusso and rang the bell.

They were just finishing supper and were surprised to see me again so soon. I didn't say a word.

"*Buona sera, Tiziana. Buona sera a tutti.* Uh, please excuse the disturbance, but Alan has had an accident. Could you give us a hand?"

Tiziana's husband came along with us, and Giovanni followed in the big tractor. When we all got back to the scene of the "incident," we surveyed the situation: the upturned car lay like a big dead insect among the vines: "*Cristo, Cristo, Cristo! Cos'é successo!*" exclaimed Giovanni. It was more a statement than a question and I remained sheepishly silent.

We ran thick canvas straps from the tractor over and under the car and attached them to the axles. Then, as Giovanni slowly backed up in the tractor, the three of us pushed on the far side to right the car and drag it up onto the road. We towed it back to the cantina with me behind the wheel miserably, guiltily steering the dirty, wet, leaf-covered wreck.

Ivana appeared, alarmed that her brother had not shown up for dinner, and I had to explain the whole unexplainable thing to her. In a sad twist of fate, I was supposed to give the car back to Renata that evening (her son was coming back the next day), so I had to tell her what happened as well. I apologized profusely and promised that I would have it fixed good-as-new as quickly as possible and offered to rent her a car in the meantime.

And I did. It took over two weeks and cost over 2,000 euros—everything I had earned during the entire harvest plus some. But it took much longer than that for me to get over the stupid accident.

Suddenly, it seemed as if my year-long love affair—with Ivana, with Italy, with the vine—had taken a drastic turn and led me into disaster instead of bliss. Fabrizio joked that I was so drawn to the vineyard I had driven right into it. I know he was just trying to make light of the situation, but in a way, I thought, he was absolutely right. "*Beh,* you might want to try using a tractor next time," he suggested, mustering a weak smile and patting me on the back.

PART V

DORMANCY

34 · WINTER WANING

There's nothing serious in mortality.
All is but toys; renown and grace is dead,
The wine of life is drawn, and the mere lees
Is left this vault to brag of.
—William Shakespeare, *Macbeth*

THE *PERMESSO* BATTLE and the car accident set the stage for the changing season, like a gray scrim dropping down from behind a curtain to darken the sunny set of spring and summer.

The days grow ever shorter; the air cold, damp, and heavy. Everything is quiet, somber, submerged: everything is dead, or perhaps only sleeping very, very deeply, but it's hard to tell which. I trudge through my mundane chores in the chilly cantina and trudge home through the gray light to a solitary dinner, the bare bulb hanging from the ceiling artificially obliterating the expansive darkness outside.

Sometimes—more and more, in fact—there is nothing for me to do in the cantina and I don't trudge out at all.

The late fall/early winter is a quiet, interim period in the wine-making world: harvest is over but pruning has not yet begun; fermentation is finished and the wine safely tucked in for a long, fruitful rest. Most of the major cleaning and maintenance projects have taken place. There is some bottling to do, an occasional *travaso* or other small task where I can be of help. But the pressure is off, and so, much of the time, am I.

Changes in the natural world around me only seem to underscore my internal gloom. Winter is a waning: the vines turn brown, then shrivel and drop their leaves; daylight shrinks, a cold chill descends. Everything contracts. People retreat indoors near woodstoves. Heavy coats and sweaters come out of storage; thoughts turn inward.

At the end of October when most Americans are busy putting finishing touches on Halloween costumes, people in Italy pack into cars and leave town, clogging the major *autostradas*. Like butterflies heading to the beach each year, people go back to their ancestral home, the place where they came from, where their loved ones are buried. The few that have no loved ones just go on vacation.

The first of November is a holiday called *Ogni Santi* (All Saints' Day) and that is followed by the *Giorno dei Morti* (Day of the Dead). There is no music or dancing on this festival, but it is not a totally somber one either. At the onset of winter's dark chill, people remember those who are no longer with them. In villages throughout Italy, Mass is celebrated right in the cemetery and people arrive early and linger after the service to spend time with their departed loved ones.

Ivana too makes a visit to the cemetery but prefers to skip the official ceremony.

"People put on their best fur coats even if it's still warm, they have their hair done, get dressed up, and go there to see and be seen; who's wearing what, who's with who, who has the biggest flowers for their grave—I don't like that." She goes the day before the actual holiday when it is quieter. And I go along with her.

Cemeteries in Italy are not creepy, out-of-the-way places. Usually they are just outside the city limits, readily accessible yet separate from everyday life, like a kind of small satellite community or suburb, a *frazione* of the dead.

Some people are interred in the ground as in the States or in family crypts that resemble little cement houses. But most are buried in a wall of tombs four or five high (there is a tall ladder on wheels to access the upper ones). And most of these tombs follow the same basic format: each space is like a long drawer about two feet high by three feet wide fronted by a rectangular marble slab, and on this slab is a picture of the person interred within. (How, I wonder, do people choose these pictures? Who does the choosing? Are they taken way in advance especially for this purpose? Or dug out after the fact from some family photo album?) Also attached to the slab is a small vase for flowers, usually plastic or silk (each bunch is different and an entire wall of them creates a colorful collage), and across the front are brass letters with the name of the person and the essential dates of birth and death. Sometimes there is also a little additional information such as married or maiden names, honorary titles (doctor, cavalier, military rank), or some clue about the circumstances of their demise ("Victim of a tragic automobile accident," "Taken from us as an innocent babe," etc.).

Many tombs also have a little electric votive light off to one side. This light, a sort of modern-day eternal flame, costs a bit more to maintain but many obviously feel the extra expense is justified because there are lots of them. They give the place an element of warmth and cheeriness, a sense of being "occupied" or "open for business," like hundreds of illuminated NO VACANCY signs. At night, the walls glow with these little flickering votives, making it seem as if there's an exclusive party going on, a party to which we, the living, are not invited.

Typically, cemeteries consist of three such walls of tombs arranged at right angles to one another, connected by a gated wall in front to form a square (larger cemeteries may have several adjacent squares or subdivisions within a square). Walking in through the gates, which often have some sort of benediction written over the entryway, you feel you are entering a protected enclosure. Inside, the atmosphere is peaceful and reflective, like that of a cloister. Everything faces inward, and the sound of gravel crunching underfoot echoes against the silence of eternity. But the silence, oddly enough, doesn't seem heavy or absolute. These walls are populated—it feels not unlike a city block of high-rise apartment buildings—and the inhabitants of this little community seem to emit a constant soft chatter of the past. Yes, everybody is dead but the overall sensation is of life, of a sleeping village all around you. The mere act of pausing in front of a marble slab to look and read seems to flick on a light or open a door on a life that was. The person inside stares out at you proudly, obstinately, through the photograph.

"I lived," they seem to say. "I loved and was loved. I had a life with joys and pains, a story like all the others and yet special, mine alone. Now I am here, dead—as you too will be one day—but I am not forgotten; I lived. And my death, my absence, leaves—whether you realize it or not—a hole in your world that no one else will ever be able to fill."

★   ★   ★

"Where are your people buried?" Ivana asks as we walk through the open gateway into the Monforte cemetery. Inside she automatically crosses herself and does a slight curtsy.

"I don't know," I reply softly.

"What do you mean you don't know? Don't you ever go to visit them?"

"Well, I hate to say it but I really don't know. My father is still alive. My mother was cremated according to her wishes and her ashes were scattered on the beach at a favorite family vacation spot so they are gone into the sea now. As for the rest of my relatives, I really don't know. I'm sure my grandparents are buried somewhere in Chicago, but I don't know where. We never went there when I was young. Most people consider it morbid. Or maybe my parents just didn't care. And my uncle, he died about ten years ago, but I hadn't seen him since I was a kid; I was far away and wasn't even invited to the funeral."

As we turn a corner filled with maintenance paraphernalia—ladders, brooms, watering pails and glass vases near a spigot; old flower pots, dried-out plants, and a lidded garbage can—Ivana shrugs her shoulders and slowly shakes her head: "How strange you are, you Americans!"

To that last statement I can offer no defense; it is strange, this disconnection from the dead and from the past that I and most other people I know in America have, an odd and unfortunate dislocation. But I don't even need to try to explain it, for we have come to a stop in front of a slab on the second tier of tombs and so has our conversation.

And I don't need to look at the brass-lettered name to know where we are: staring out at us at nearly eye level is the spitting image of Ivana! The same narrow face, the same dark eyebrows, thick black hair, and the same look of gravitas—the resemblance is immediate and unmistakable.

FRANCESCO MASCARELLO

★ 1925

† 1980

HUSBAND, FATHER, WINEMAKER

Ivana takes two steps forward, dusts off the likeness, and kisses it. She steps back, looking her father in the eye, and silently talks to him while the birds chirp and the cool winter sun twitters in and out between the clouds. I stand respectfully toward the back, not wanting to interfere.

Moments pass, timeless moments, moments of calm communion and exchange. Then Ivana crosses herself again and once more she makes a slight curtsy. She affectionately touches the picture like someone familiarly caressing a cherished face, kisses it, and crosses herself one more time.

Then we go to work, as if cleaning the house before a party: Ivana wipes the dust off the marble slab and the brass letters and vase. We go to the spigot and get a pail of water. The dirty plastic flowers are removed, water is poured into the vase and fresh flowers added in their place. "They'll only last a few days and they're expensive, especially this time of year, but that's how it is. They're *ladri*, those florists—they make a real killing around the Day of the Dead!" A potted flowering plant is placed at the foot of the wall, and bits of gravel, leaves, and twigs swept off the sidewalk in front. Ivana pauses; she touches and kisses the picture one last time. She backs away slowly, lingering, then turns, and we move on.

She was only nine years old when her father died; nearly a quarter of a century has passed. Yet he is still with her in some way, still part of her life.

"Oh yes," she says. "I come here regularly. I talk to him, I ask him for advice. And he helps me. I think he is looking out for me. I miss him terribly; nothing will bring him back. But I feel better after visiting him here."

We stroll along the gravel streets of this peaceful village hand in hand. As we pass the walls, I see a horizontal roster of familiar-sounding names from the world of wine go by—Conterno, Rinaldi, Valentino, Clerico, Mascarello—and a constant stream of faces, old and young, each a snapshot of a personal world that was, a world that for some perhaps still is.

We pause before another couple of slabs, this time on the third level, and Ivana reaches up to touch the lower corner of each in salutation. "These are my grandparents," she says as if introducing me. "Hello, pleased to meet you," I think to myself.

We make a number of additional stops to visit her great-grandparents, whom she never even knew (they died well over a century ago), and other people, now deceased, who touched her life in some way: a priest who was Ivana's teacher in Monforte, a great-uncle who hunted and always smelled of dead animals, and a youth of about eighteen who tragically took his own life.

I envy Ivana's connection to this village of the dead and to the past, populated by people who once were. Ironically, it seems to help anchor her to the present, to this place in the world and the reality of her life, providing depth and stability like the deep roots of a vine. In comparison, I feel alone and adrift, disconnected from those who have preceded me.

As we stroll out the gates of the cemetery, back to the world of the living, the soft chatter recedes, evening falls, and the little lights come on: a nocturnal celebration is about to begin, one to which we are not invited—not yet, anyway.

Outside the cemetery, the sound of our feet crunching fallen leaves blots out the echoing murmurs of the dead. At first, the world outside the cemetery walls seems hollow and deserted. We crave real voices and the company of the living. Plus we're hungry.

"Andiamo da Gemma?" I ask, breaking the silence.

"But did you read my mind?" replies Ivana. "I was thinking the same thing; I'm starving!"

We jump in the car, drive down to Monforte, bear left at the piazza, and head up through the twisting twilight-shrouded hills toward the little village of Roddino.

We leave the car at the base of the wide stairway beneath an imposing church, which is closed, lifeless, and eerily illuminated. The whole town, in fact, seems deader than the cemetery: there is nobody on the streets and the houses, shuttered tight with nary a splinter of light peeking out, resemble anonymous crypts enlivened by neither eternal flame nor posted picture.

Undeterred, we walk up the street below the church and turn off at an inconspicuous little house. We go through a small gate, down some stairs, and up to a side door. There is a momentary pause: "Is this the right place?" I always ask myself. But I know it is. I turn the knob and push open the door; we step inside and there we are—solidly, indubitably back in the warm, sweaty, smoky, cacophonous world of the living.

On the right side is a long bar with a sparse collection of half-filled dusty bottles and a TV in the upper corner that is always on (but the sound is always not); on the left is a wooden phone booth and a few tables strewn with wrinkled newspapers and old magazines. Past the bar is a square room with several slim metal pillars down the middle. On the right, most of the tables are occupied by men playing cards and smoking. Each time someone new comes in they look up from their game to see who it is. And I look at them too: there is a little guy with disproportionately large head, deep-lined face, and muscular shoulders that I have seen each and every time I've been here. I guess he has seen me too because we give each other a slight nod.

On the left side is the dining room, which is usually filled with a boisterous collection of people; couples, families, or groups of rowdy men who always seem to break into song at the end of the evening. Local chefs come on their night off just to eat and relax, and winemakers come here too when they don't need to try to impress prospective customers.

The faded walls are scattered with an odd collection of drawings and photographs, some of which hang askew, and there's a potbellied stove in the center of the room which radiates heat during cold weather (when not in use, the stove offers yet another place to rest wine bottles or trays). On a table toward the rear is a wooden board covered with flour, a ball of dough, and recently cut *tajarin*.

As we make our way to a table, Gemma waddles up to greet us. She is late fifty-something, soft and fleshy, but strong. A stained apron is tied around her thick waist, and flour flecks her wispy hair. She has a gentle, quiet sing-song voice, and her big eyes seem tinged with a suggestion of some old sadness. She is a gracious host, a kind of benevolent every-mother, despite the fact that she is shy and reserved, generally preferring to hang back on the periphery than bask in the limelight.

"*Ah, ciao*. Nice to see you. *Allora? Come va?*"

"We are fine, Gemma. How are you?"

"*Eh beh,* how am I? As I always am: tired. I'm happy; business is good, *grazie a Dio*. But I'm tired. The old people at lunch, the young people at dinner, the cardplayers all the time; I can't ever close. Where would they go? And I have no one to help. Sometimes I ask a neighbor to lend a hand"—at that moment a little old woman who must be near ninety zooms by with a platter full of *agnolotti*—"but that's it. And anyway, even if I had help, where would they work?"

With eyebrows raised and lips curled into a scowl, she indicates the kitchen at the rear. It is about the size of a closet; there's a single oven with four little burners, a sink, a home-size refrigerator, and a small counter. Food at various stages of readiness is everywhere—in strainers, pots, roasting pans, bowls, and on cutting boards—leaving just barely enough room for Gemma and her dishwasher-helper to squeeze in.

"*Beh,* maybe one day I'll get a nice new place, spacious and clean with a separate room for the cardplayers. *Magari!*"

I hope not, I think to myself as we sit down. I sympathize with Gemma and want her to be happy. But there are so few places like this left: a real old-fashioned *osteria*, a true *trattoria*. Actually, this place is neither (it's not even really a restaurant), but rather what is known as a *circolo*, a private club that technically requires annual membership—but the feeling of cozy authenticity is the same; a simple, genuine place where you feel comfortable and well taken care of. And well fed.

Here there's no menu, and the meal is almost always the same. When we sit down an open bottle of dolcetto is placed on the table along with a bottle of water (we use the same stemless glass for both). Then a wooden cutting board with a salami *cotto* and *crudo* (both made by a local farmer and absolutely delicious) is brought out with a big sharp knife. This is dangerous: you have to resist eating more of the *pan e salam'* than you should and filling up right there. Then the appetizers start to arrive: Carne Cruda, Vitello Tonnato, a yellowy-rich *Insalata russa*, followed by the *primi*, handmade *tajarin* with thick meat sauce or pinched ravioli (we ask for a taste of both), served from a large platter. After that, if you still have room, there will be a few main courses to choose from, rabbit or *brasato* or *bollito*, and then a strudel or meringue.

Simple, typical, predictable, but thoroughly satisfying and not too expensive either, dining *da Gemma* is certainly not what one would call refined or sophisticated. Also, as Ivana points out, the place is undeniably a little *sporco* and the cigarette smoke can sometimes be annoying. But I like it the way it is.

There are many restaurants around here; new fancy restaurants with elegant food served on large porcelain plates or old places that have been recently renovated and updated. I understand the practical benefits of modernization. In any event, things change. And sometimes the food is just as good as it ever was. But often, something is missing; few of these new improved places leave me with the good feeling that a visit to Gemma's does.

Sure, I want Gemma to be happy. But when she gets her new place, I know I will be a little sorry too.

Once we have finished eating and drinking and listening to the men sing and watching the regulars play cards, we thank Gemma for her hospitality, bundle ourselves into the car, and head back down to Castiglione for an impromptu little party of our own. When we get home, we rush upstairs, throw off our clothes, dive in between the chilly sheets, and revel in another of the activities that, like going out to dinner, is the sole prerogative of the living.

· RECIPE #23 ·

VITELLO TONNATO

Sliced Veal with Tuna Mayonnaise, adapted from Osteria Gemma, Roddino

I. PREPARE THE TUNA-CAPER MAYONNAISE

THIS RECIPE MAY be prepared a day or so in advance. You can make the mayonnaise by hand in a mixing bowl with a whisk or by using a food processor.

INGREDIENTS

3 egg yolks   *1 heaping tablespoon small*
Juice of ½ lemon   *capers* (see note below)
½ teaspoon white wine vinegar   *½ teaspoon white wine*
½ cup extra-virgin olive oil   *Salt and pepper to taste*
1 three-ounce can of high-quality
    tuna fish, drained of oil

1. Put the egg yolks in a stainless steel or ceramic mixing bowl (or in the bowl of a food processor fitted with the steel blade). Add the lemon juice and vinegar and whisk vigorously.

2. Add the olive oil in a slow, steady stream, whisking/processing continuously until a thick mayonnaise is formed.

3. Chop the tuna and capers and whisk into the mayonnaise. (If using a food processor, transfer the mayonnaise to a bowl. Add the capers and chop. Add the tuna and process. Return the mayonnaise to the processor bowl and pulse to combine all the ingredients.)

4. Add the white wine and mix until a creamy, spreadable mayonnaise is formed. Add salt and pepper to taste.

YIELD: MAKES ABOUT I ½ CUPS OF MAYONNAISE.

II. PREPARE THE VEAL

INGREDIENTS

1 (1-pound) piece of veal from   *1 bay leaf*
    the leg, preferably top round   *4 black peppercorns*
2 ribs of celery   *½ teaspoon white wine vinegar*
1 onion, peeled and quartered

1. Remove any fat or cartilage from the meat, and, using butcher twine, tie firmly to create an even cylindrical shape. (You may have your butcher do this for you.)

2. Cook the meat *(see note below)*: Place the celery, onion, bay leaf, and peppercorns in a pot large enough to accommodate the meat. Fill

it with cold water and place over a medium flame. When the water heats up (but is not yet boiling), add the vinegar and the meat. (There should be enough liquid to completely cover the meat.)

3. When it comes to the boil, lower the heat and simmer gently until the meat is cooked through (approximately 30 minutes).

4. Turn off heat and let the meat cool in the liquid. Store the meat in the liquid in the fridge until ready to use.

III. SERVE

| Salt | 1 teaspoon fresh chopped |
| 1 teaspoon capers | parsley |

1. Slice the veal as thin as possible and arrange in a single layer on individual plates or a large serving platter.

2. Sprinkle the meat lightly with salt, spoon the mayonnaise over the meat, and spread to cover evenly. Sprinkle the capers and parsley over the top and serve.

YIELD: SERVES 6 TO 8 AS AN APPETIZER OR 2 AS A LIGHT MAIN COURSE.

NOTE: I prefer the capers packed in sea salt from the volcanic island of Pantelleria off the coast of Sicily. Salted capers must be well rinsed before using.

Gemma boils the veal in the traditional way. Another method of preparing it (and one I myself generally prefer) is to season the veal with salt, pepper, and olive oil and roast it in a preheated 400-degree oven for about 20 minutes. The advantage to this is that the meat, when properly cooked, retains a nice rosy color and moistness.

35 · CYCLES OLD AND NEW

Beauty and the Beast

FABRIZIO'S VINEYARD IS a vineyard in transition. Actually, it's in the midst of several transitional cycles.

There's the annual cycle that I experienced when I first started coming here; the dormancy of the vines, the pruning, training, growth and grooming, harvest, and dormancy again. This happens in every vineyard every year.

Then there's the life cycle. This takes place much more slowly, perhaps every twenty or thirty or forty years or more. Vineyards, like people, have a life; they are born, mature, grow old, and eventually die. With good pruning, limited yields, and continuous replacement of individual damaged or diseased plants, a vineyard can theoretically last indefinitely. But this is usually not the case. At some point, for one reason or another, a vineyard gets tired, reaches the end of its productive life and needs to be ripped out and replanted from scratch. This is the current state of Le Munie.

There's also a historical cycle. This is the slowest one. Actually, it's not even really a cycle since it never repeats or comes full circle; it's more linear, like an evolutionary time line or a continuously ascending spiral. It began far away at the dawn of civilization when people first transplanted wild vines to the base of trees and began to plant primitive vineyards. It was continued by the Egyptians, and the Greeks, and the Romans, and the monks, and countless anonymous farmers throughout Piemonte, Italy, the Mediterranean, and the world, right up to the present. Le Munie is in the midst of this kind of transition too.

When I first started to work in Le Munie, the poles holding up the wires were all a jumble of old wood, there was an ancient well dug by God-only-knows-who, and a whole vineless weed-covered section that had been devastated years before by a lightning storm. The rows were arranged in something of a hodgepodge fashion and some of them were barely wide enough to walk down, much less drive a tractor through.

Now it is all bare.

When Fabrizio replants, he will use cement posts that stand straight and never rot; he will get rootstock from a couple of proven nebbiolo clones and create even, well-spaced rows from one end of the vineyard to the other. Le Munie will get a face-lift, dress herself with clean, modern-day accouterments, and blend in with her up-to-date neighbors. Most of them, anyway.

Most vineyards in the Langhe have already gone through this kind of transition while others like Le Munie are in the middle of it. But you can still come across a few that linger obstinately behind, standing out distinctly from the straight, even rows around them. If you happen to come upon one of these, stop and look at it, take a picture, for it will not be there for long. This is one of a nearly extinct and rapidly disappearing breed, the Traditional Piemontese Vineyard.

These old-fashioned vineyards consist of thick wooden headposts shooting up out of the ground at the beginning and end of each row. In between, every couple of meters or so, are thinner wooden poles called *pali di legno*, which are cut from the nearby forests. Unlike the uniform modern cement or metal posts, these sticks go bending and twisting up at all different heights; some, in fact, tower high above the upper wire. There is a reason for this. The end of the wood pole sticking into the ground is subject to rot; during the first spring cleaning *(scalusé)*, loose or rotten ones must be pulled up and the end cleaned and sharpened to form a firm, new point. Clever farmers left the poles extra-long so they could be whittled down year after year as necessary without needing to be completely replaced.

Wire cords running the length of the row are tied, nailed, or stapled onto these wooden poles. The *pali di legno* in turn are interspersed with reed canes *(canne)*, one for each vine, that reach to the level of the top wire. It is still common to see stands of exotic-looking bamboo-like cane growing next to vineyards, though they too are becoming increasingly obsolete, an unnecessary waste of space except to the few people who still use them. (Plastic canes or metal rods can be used instead, and, with the new solid metal or cement posts and high-tension wires, thin canes in between the poles are all but unnecessary anyway.)

Also growing alongside many vineyards are little trees called *salice,* a type of dwarf willow. They have narrow straight trunks about four feet tall and branches sticking out in all directions like the head of a sparkler. The supple tips of these long, narrow branches were trimmed to create ties, called *gurates,* used to secure the vines to the wires and canes. Many of these trees still exist, though fewer and fewer people use willow tips to tie vines, most opting for wire, plastic, or string instead.

In the late spring and summer when the tips of the vines grow beyond the height of the upper wire and start to sag over, instead of being cut they are twisted dreadlock-style and folded one over the other, creating a wavelike crest along the top of the row.

Sprinkled throughout these old vineyards, articulating the landscape as if popping their heads out from among the vines, are little structures—small shelters, actually, one or two stories high—called *ciabot.* Usually made of timber and stone, they range in size and style from rustic sheds to small houses; some of the more elaborate ones even have fireplaces and balconies. These *ciabot* are remnants of a time before tractors and cars when people had to walk or ride an animal to get to the vineyards. During harvesttime or bad weather, it wasn't always possible to get home, and people needed shelter, sometimes for several days on end. Nowadays, most of the remaining *ciabot* are abandoned, used only for storage or trash if they are used at all. Although rickety and decrepit, many of them are lovely old buildings that stubbornly retain an aura of character and individuality and lend a distinctively human element—that of a (former) shelter or habitation—to the vegetative surroundings.

In plots that were too small to merit an actual *ciabot* (or vineyards that were close enough to home not to need one), farmers constructed wooden arbors covered with vines—often sweet moscato—which provided respite from the summer sun during the afternoon break when it was too hot to work. Often, near a *ciabot* or arbor you find an old well made of brick or stone. Sometimes the openings are covered with an iron lid or wooden closure to prevent people from falling in, sometimes not.

In these older vineyards you may still come across different types of vines—nebbiolo, barbera, dolcetto—planted side by side. The reason for this is that the small plots were worked by the same farmer year after year; he knew the subtle variations of soil in his bit of land and planted different types of vines to take advantage of them. Sometimes these different grapes were picked and vinified separately, requiring numerous passes through the vines during the long harvest season; sometimes they were simply mixed together. You may even see vineyards with vegetables planted in between the rows of vines as was the old practice. The primary reason was not to waste space; some feel that the intermingling of agricultures (nearly impossible in a modern commercial vineyard) also benefited the grapes by preventing disease and strengthening the soil.

These old vineyards are strange and lovely to behold; they are like living relics, strikingly appropriate expressions of this region from a very different time and perspective. They seem to spring right up from and blend in with the landscape, a perfect collaboration of nature and man. Made of local, organic materials, they are ever-changing, living, breathing entities, much like the vines themselves. Each one is irregular, handcrafted, and unique, a reflection perhaps of the person who made it.

I think these old vineyards are beautiful. But, like an old farmer with his wrinkled face and slurred speech (the image of Bastian at the harvest table pops into my mind), their days are numbered. They are rapidly disappearing; in a handful of years they will surely all be gone, and with them a whole world of knowledge developed over time and passed down through generations will be gone too.

Old-fashioned vineyards may be attractive but they are also more difficult to work than the new ones. When you yank the dried branches down during the first pruning, half of the canes come down with them. And because of the tall, uneven poles it's impossible to use a tractor attachment to trim the vine tips; each needs to be cut or folded over by hand. All these things take time, a commodity most people don't have enough of anymore. The time Fabrizio will save with his new vineyard is precious time he can spend with his family. Hard to argue with that, especially when modern vineyards don't necessarily diminish the character or quality of the grapes that are produced in them (in fact, some would argue the opposite). And harder still when the European Community offers financial incentives to farmers who "upgrade" and modernize their vineyards.

I understand all this and know that this transformation is an inevitable one, an irresistible pull of progress. I'd probably upgrade myself if I had an old vineyard. But it makes me appreciate all the more the ones that are still here. I will miss them when they are gone. And, as time marches on with an increasingly quicker and more efficient step, crushing these old vineyards underfoot like a big steamroller smoothing out all the imperfections and irregularities, I think that something will be missing from the landscape of the Langhe, just like the old well, a lovely beauty mark just above the *vita* (waistline) of the vineyard, is now missing from Le Munie.

You can see it from my balcony as well as from many other parts of the Barolo zone: a huge blank spot sticking starkly out from the pastoral landscape. It's a vineyard, a vineyard at the same stage as Fabrizio's, getting ready to be replanted in the spring. But this one is different. Compared to the seductive vixen of Le Munie, this one is a big, muscular guy in drag, or the Bride of Frankenstein on steroids. Either way, it's not what it appears to be. And it's definitely a *vigneto*.

They've been working on it almost as long as I've been here. For over a year the spot has been a beehive of activity with four or five steam shovels crawling up and down, dump trucks hauling dirt, and a large crew of workers swarming over it like ants devouring a dead worm.

It used to be two gentle hills, like soft, round little breasts, one slightly bigger than the other. But the space between them has been filled in, the two hills transformed into one vast slope. The very facade of the earth has been changed, the rolling contour smoothed out into a flat southeast-facing wall at an even grade as if by an immense trowel over wet concrete.

This is the antithesis of the old, traditional vineyard crafted by hand in harmony with nature. I understand the rationale behind it: one continuous hillside will be much easier to work than two rolling hills. This "sex-change" of the landscape will save time and money; efficiency will be increased as will usable space (the owners purportedly maintain they will lose space by eliminating the crevices of the two hills, a sacrifice in the name of "quality," but I don't believe it). The new vineyard will be more uniform and more productive.

I cannot dispute these arguments. Even many local farmers look at the altered hillside with admiration and a touch of envy; perhaps they would do the same thing if they could. Nevertheless, it doesn't seem right to me.

The former little breasts were perfectly good vineyard sites as they were, quite prestigiously located, in fact. This alteration strikes me as an unnecessary surgical procedure. In any case, it just doesn't seem right to manipulate nature like that, especially in order to produce a thoroughbred wine like Barolo so closely tied to place, and especially in such a tightly controlled and regulated *denominazione* as this one. By law, to make Barolo, you can only use grapes grown within a small delineated area, on slopes with certain exposures with limited yields per hectare, and you can't irrigate (even in the driest summer in recent history) as it alters the natural factors that determine the character of the grapes. Why then is it okay to completely alter the natural profile of the earth in which the grapes are grown? (And where did all that dirt come from anyway? Was it DOC dirt from within the Barolo zone?) A project like this would have been unfeasible—even unimaginable—a generation ago. With today's technology, couldn't the whole unique mosaic of the Langhe with its many diverse microclimates, molded over eons through an irreproducible series of geologic and climatic events, be smoothed out and rebuilt into a series of long, uniform southwest-facing slopes? Aren't there any regulations about that?

In any case, the project has been carried out, the alteration has been made. It took thousands of years to sculpt those lovely little breasts in perfect synchrony with the surrounding landscape; how pompous of someone, a winemaker no less, to think it could be improved on. But nature will be the final judge and if it doesn't approve of the changes, it will surely sweep away the fill and return the earth to how it was. Only time will tell.

Old versus new is a heated topic when it comes to Barolo, and the kind of vineyard the grapes are grown in is just the tip of the iceberg. Most of the fuss concerns what goes into the bottle and what goes on in the cantina to get it there. "Old-style" vs. "new-style," "traditional" vs. "modern," "typical" vs. "international"—seems black and white on the surface, and many wine lovers maintain an avid allegiance to one or the other.

Perhaps these *were* valid, relevant distinctions a couple decades ago when the wine boom was in full swing, when Piemonte had just ventured out into the world, and a whole new generation of open-minded kids was taking over the reins from their predecessors. Certainly, if you compare a typical Barolo of thirty years ago—say 1971—with one of 2000 (another highly touted vintage) certain general differences would be immediately evident. One of them, obviously, is age; a mature wine, like a mature person, is quite different from a young one. Bottle age aside, the older wine would probably be lighter in color, leaner and less unctuous than the younger one (greater yield of grapes per vine, thus less concentration; lighter maceration due to the absence of mechanical paddles or plungers). The tannins in the '71, despite its age, would probably be harder and drier, having come primarily from the tough tannins of skins and stems during a long fermentation period, while those in the 2000 would probably be more soft and supple due to short, cool fermentation and maturation in sweet, toasty oak *barriques* (small 225-liter barrels) rather than large wooden casks.

It is these aging containers—small *barriques* versus large *botti*—that signal the difference for many and have become a rallying cry for one side or the other. The use of *barriques* for aging wine is another practice that was imported from France. Angelo Gaja was one of the first to use them, and a whole generation of aspiring players followed eagerly in his large footsteps. Most of the small barrels themselves are made in France, where their fabrication is an established art form. Each *tonnellerie* (as barrel-making operations are called) has its house style and tradition, gets its wood from specific forests, cuts it and dries it in its own time-honored ways, and gives its particular "toast" to the inside of the barrel.

They are beautiful objects to behold and justifiably expensive; a good, hand-crafted French barrel can easily cost close to 1,000 euros and will be effectively useful only once or twice, that is, for three to five years.[1] After that, the toast wears off, the pores in the wood get clogged up, and it becomes nothing more than a neutral storage vessel or a planter (used barrels are usually sold off at a fraction of their original cost). This expense, plus the additional labor cost of handling them, is passed on to the consumer; wines aged in *barriques* are generally more expensive than those aged in the traditional large *botti* or cement or stainless steel.

Proponents argue that the extra cost is nothing compared to the benefits; that *barriques* help produce a wine with much more class, suppleness, structure, and elegance. Detractors reply that the toasty wood overwhelms the flavor of the grapes (especially the persnickety nebbiolo) and the *terroir* they came from. They avidly protest against a sort of vinous globalization in which wine aged in *barriques* in Italy tastes exactly like wine aged in *barriques* in California or Australia or Chile, all of it made to appeal to the same international market.

I think they are both right; each side makes a compelling argument. It does seem a bit odd to require something distinctly French in order to make something distinctly Italian. Some traditionalists point their fingers derisively at the Frenchness of the terms *barrique* and *cru* as clear evidence of their foreignness and inappropriateness. But then, you could say the same about 50 percent of the Piemontese dialect as well. Should the Piemontese therefore stop speaking it? With its longstanding cultural ties and geographical proximity to France, if any region could make a claim for the right to use the *barrique* in its wine making it would be Piemonte. Someone, if they dug deep enough, could probably even find evidence of a historical precedent for the use of small barrels in Piemontese wine making, perhaps even predating their use in France.

But is historical justification really necesary? Should it not be self-evident and self-contained in the wine itself? The important thing is not so much *what* is used but *how*. Use whatever you want, the real question is what is the final outcome in the glass? Is it good? Is the wine pleasant and well balanced? Is it interesting? Does it have personality? Or is it totally anonymous? Does it speak of a particular culture, tradition, or place in the world—a *terroir*—or is it outside of place altogether, a dislocated end in itself?

As for the whole old/new debate, as far as I am concerned, it is over. The distinctions between "new-style" or "old-style" are blurred: no one wants to go back to the old days of stomping by foot[2] and making wines that were often oxidized, overly tannic, or "dirty" due to poor sanitation in the winery or total lack of control of the vinification process; today, even an arch traditionalist reduces his yield in the vineyard, keeps his cantina spotlessly clean, has samples chemically analyzed (even if only to confirm what he already knows), and inoculates his must with selected yeasts to start fermentation, all benchmarks of "modern" wine making. At the same time, ask someone like Marco Parusso, a Young Turk of the new generation who uses only 100 percent new French *barriques* for his Barolos, what kind of winemaker he is:

"I'm a traditionalist," he replies, without missing a beat. "My goal is to express the unique characteristics of our individual *terroirs,* and there is no better way to allow that to come through than with a high-quality new French barrel; it is neutral, pure—it lets the wine breathe."

And this is as it should be. Things change. Tastes change. Barolo, after all, started out as a sweet wine (and if it were not for the influence of the French a mere hundred and fifty years ago, it might still be). How far back do you need to go to make a "traditional" one? And if you don't go all the way back, aren't you just picking an arbitrary point along the continuum of change and labeling it "traditional"? *Traditional* to whom?

It's time to let the irrelevant classifications of old/new rest in peace. "Good" and "bad" are the only classifications that really matter. There is room for many different interpretations of Barolo, *valid* different interpretations, as long as they are done well and honestly (rather big qualifiers, those), as long as (and to the extent that) they express the unique qualities of this particular grape and this place and the cumulative histories of the people who have long engaged themselves with the vine and the earth here. Winemakers, each drawing on a knowledge of the individual *terroir* and his own background and philosophy, can decide for themselves what style of wine to make and how best to make it. Consumers, too, can decide for themselves what style of wine to drink. And if some consumers have a natural preference for wines that change with each vintage, that show a certain earthiness, rusticity, and cantankerous complexity, a lack of polished finesse, a kind of rough tannic dryness and limpidity that express well-tended vineyards but also seem to recall the directness of a time gone by, well, that's okay, too.

36 · LOCAL LANDMARKS
The Cedar and the Cube

JUST OUTSIDE MY apartment, across the road and behind the public scale where we weighed the harvested grapes of Le Munie, is a broad terrace looking out over the vineyards to the west, north, and south, offering a 180-degree view of the surrounding countryside.

Directly below, the vines fan out in a big open semicircle resembling some ancient earthen amphitheater. Curving off to the west is the vineyard called Monprivato with the Ca' Nova up on the right above the Alba-Monforte road and the Ca' d' Morissio (both uninhabited houses) down along the shortcut through the vineyards. To the right of this asphalt road a small corner of the Castiglione cemetery peeks out from behind the vines, while on the opposite side of the valley the white Cascina Brunella stands sentinel. Straight ahead, a large, sweeping ridge dominates the horizon. In the middle of this graceful arc of land where the summit should be, is instead a dip that forms two humps; on the left side, a group of antennas sit atop the hill like boat masts marooned on a sandbar, while on the right the lofty village of La Morra caps the crest, its towered belvedere winking directly across the valley. Slightly to the right of and beneath La Morra, the twin hamlets of Annunziata and Santa Maria huddle amid the hills while higher up and farther to the right, the villages of Roddi and, beyond that, Verduno fade off into the distance.

From the terrace railing, one looks over the undulating hills like the captain of an ocean liner looking out over the cool rolling waves, one upon another off into the west and north for as far as the eye can see. In the spring and summer, they resemble folds of a lush green velour fabric, while in winter, the snow-covered ripples remind me of a white, frozen desert.

This sweeping panorama is articulated by a couple of landmarks that pop up from various vantage points throughout the area, making them points of reference and orientation.

Over to the right, just behind and slightly higher than the two-hills-made-one is another pointed hill, and at the very top is a tree. But not just any tree; it is a marvelous, majestic old cedar of Lebanon, "proud and lofty . . . lifted up and high."[1] It springs right up from the center of the peak and spreads out over the top like an expansive green fountain frozen in mid-air. Vines rise up the hill on all sides but stop a respectful distance from the tree, creating a kind of aureole around it. Here people come to picnic during the day and couples come to make love at night. From underneath, the tree seems like a giant octopus stretching its long arms out in all directions as if it were trying to grab a distant castle or village, while higher up the branches resemble the arched vaults of a Gothic cathedral. From a distance it looks like some sort of crown capping the crest of the hill; if you didn't know, you might not even realize it was a tree. You can see it sitting there atop the hill from far and wide, regally surveying its domain. But the tree doesn't tower over or dominate the landscape. Rather, it is embraced by it. Like Bartolo in the next village over, the tree is a passive sovereign. It keeps constant watch on what goes on around it as it has for well over a hundred years. It surely sees what has taken place beneath it right next door, but doesn't interfere, doesn't even deign to comment. It is not necessary; its mute presence is sufficient.

This tree didn't just happen to pop up out of the top of this hill; it was planted here in 1856 by a Falletti—Constanzo of nearby La Morra—to celebrate his marriage to Eulalia Chiesa. Thus, like the adjacent earth-moving project, this ancient cedar on the hill is a man-made creation. But the motivation was not higher production or increased efficiency or greater revenue, but love. And the tree has weathered the test of time; it has survived. It has sunk its thick roots into the earth and thrived. It has blended into the landscape and been enveloped by it in turn; it has withstood lightning strikes and floods, wars, and the pressures of development. Through it all, it is still here, living and breathing, hugging the crest of the hill, watching over the surrounding countryside, and being watched by all; offering solace to those who come to sit beneath it and a poignant example to those who care to heed it.

Of *this* modification of the landscape nature obviously approves.

Over to the left, on another *bricco* (hill) stretching up above the Rocche vineyard right here in Castiglione, is another landmark jutting up out of the top of a hill. It is a cube (yes, a cube!), *cubo* in Italian, though some people refer to it as the *incubo,* the nightmare. And, like the big bare hillside across the valley, the cube is not what it might at first appear to be—a stray piece of some outdated modern exhibition hall gone bust, waiting to be carted away and melted down into reusable materials. No, this is a permanent installation.

Unlike either the fabricated vineyard or the great cedar, this cube is made of unnatural materials (or at least materials so far transformed from their natural state that the connection no longer applies), of hard steel and cold glass; its sharp edges cut into the surrounding landscape like a butcher knife while its pointed corners puncture the sky. This cube doesn't do anything; it doesn't even stand straight as most cubes do, but shoots up at an irritating diagonal, as if thumbing its nose at everything and everyone around it.[2] Even worse, it is as big as a house, illuminated at night, and strategically located on a prominent hill so as to be seen from almost everywhere.

I suspect its advocates (of which there must be at least two—the one who designed it and the one who paid for it) would probably argue that the cube's jarring presence actually enhances its surroundings; that it is a "site-specific structure" that "engages" its environment, creates a "dynamic dialogue" with its setting, attracts attention to this hill in the middle of Barolo like a beacon, and confers status on it like the rakishly tilted square cap on a graduate's head.

But the fact is, it is just ugly. It doesn't commune with its environment, it clashes with it, and it loses the battle hands down. It doesn't add anything to this gorgeous hilltop other than an eyesore (ironic since it was probably paid for in part with profits from the vineyards around it) and doesn't make any statement other than to reaffirm the folly of man. The cedar is a beautiful expression of love, still living in harmony with its surroundings after one and a half centuries; the altered vineyard, a twisted attempt to improve nature so as to better exploit its bounty. But this structure is merely a monument to the ego. It serves no real purpose (other than perhaps self-promotion and self-exultation) and inflicts itself on everyone else who has to look at it, staining the landscape like a mustache on the *Mona Lisa*.

Maybe I'm totally wrong. Perhaps in a hundred years (if it lasts that long) the cube *will* have blended in with the environment (or, more likely, the environment will have been so ravaged as to blend in with the cube). Maybe I'm just an old-fashioned, closed-minded stick-in-the-mud. Maybe one day the cube will turn out to be the work of a visionary pioneer, forerunner of a revolutionary "new reality," and I was just too stupid to see it. Certainly this is possible; it has happened with many great works of art in the past. But I don't think this is one of them. Once again, only time will tell.

37 · DESTINY

"HOW ABOUT YOU?" asked Ivana's mother one chilly evening after an informal supper in the Bussia. "Do you believe in *il Destino*?"

It was only the three of us, sitting around the kitchen table in the radiating warmth of the woodstove. Ivana and I were both anticipating the pleasure of being under the thick covers at Castiglione, but were not eager to venture back out into the cold to get there.

"What do you mean?" I ask, mostly stalling for time to try and formulate a reply.

"You know, that it is all written out ahead of time—your life, everything that happens to you—by the big guy up there, *il Gran Signore*," she says, impressively rolling the *r* of *gran* and pointing a knobby finger upward. "Some people believe we are dealt our hand right when we're born; you can fight it if you want to but it's pointless—in the end all you can do is play it out. She [indicating Ivana] believes in *il Destino*. I guess I do too. How about you?"

We had been talking about their next-door neighbors. On one side, the son contracted a strange illness as a kid and has been confined to a wheelchair, barely able to speak, ever since. On the other side, the guy had a tractor accident: he hit an invisible rock below the surface, went flying forward off the tractor, and got run over by it, leaving him hobbling on two canes and in leg braces for the rest of his life. *Il Destino*.

I think about Ivana.

Her father died when she was young. Had he lived, everything would surely have been different. He would likely have become a prominent winemaker, one of the big shots, riding the crest of the international wine wave; his *azienda* would have grown into a valuable concern, and would eventually have been transferred to his children

(Ivana would have occupied herself with marketing and public relations). But this was not to be. He never smoked and hardly drank (except for a glass of wine with his meal), but he became ill anyway and
died young. Ivana grew up in a paternal void, her father but a fond
memory on a marble slab in the Monforte cemetery. *Il Destino.*

Life goes on.

Later she fell in love with a policeman from Puglia and got married.
Had things not gone bad with her husband, Ivana would still be with
him, in a new house in Alba filled with kids, perhaps, and a holiday
bungalow by the sea. But this was not to be either; she left what little
there was behind and returned to the Bussia with nothing but the
clothes on her back. *Il Destino,* yet again.

Surely Ivana pauses every now and then to think about "what
might have been if . . ." But she doesn't dwell on it; she quickly lets it
go, returning with near total acceptance to the reality that *il Destino* has
dealt her. *"Beh, è così la vita,"* she says with a shrug and moves on.

Then I think about my own life. I have been extremely fortunate;
I've never had to go hungry, never had my life threatened, and never
had to sleep out on the street. I've had nearly unlimited possibilities.
And what have I done with them? It all seems so arbitrary: do this, do
that; go here, go there. The options seem infinite and equally viable,
yet each choice has its consequences.

Both organized religion and common wisdom say that each of us is
here for a reason, that each of us has a purpose in life, large or small; a
place in and a path through the world. But how does one find it? Is it
hidden, like treasure? Do you need to search for it and discover it, or is
it actually right in front of you, in front of each of us, and we just need
to learn how to recognize it or decipher it?

To me, it doesn't seem like the script is already written and I just have
to act it out; in fact, it doesn't seem that there is anything written at all,
other than maybe a rough sketch. I think of a recurring dream:

I find myself alone in a white room confronted with a number of
blank doors and a bunch of identical keys on a nearby table; I must pick
a key, pick a door, and on the off chance that the key fits and the door
opens, enter it, not knowing if it is the "right" one or where it is going
to take me. Each time I have this dream, I wake up, full of agitation,
without having yet made my choice.

Sometimes life seems like a big spontaneous improvisation—a matter of winging it, playing it by ear, reading aloud hastily scribbled lines before an audience of strangers without knowing what's coming up next—and there's never any going back to do it over again.

I wish there were a *Destino*, some omnipotent author/director just offstage pulling the strings. Then I wouldn't have to choose or take responsibility for the choices I have made. Then perhaps I wouldn't have the sense of wandering blindly through life, trying to feel my way along and bumping headfirst into obstacles and dead ends. But that is how, in this dark winter, it often appears to be. In any case, however much I might *not* enjoy it, I would never voluntarily surrender that illusion of control, the Choice.

Ivana's mother continues to stare at me inquisitively, waiting for an answer, while her daughter silently gets up to put another log in the sputtering *stufa*.

I think perhaps a hand of some sort *is* dealt to each individual at birth, like a set of basic character traits or potentialities. But how to play them—which cards to discard and pick up, what strategy to employ, indeed the very game itself—is up to you, along with a certain element of chance or fate or luck that is beyond your control. You arrange your cards, you play the game as best you can, you make your choices—and shit happens.

What, I wonder, was the series of events that brought me here to Italy, destiny or chance?

When the car rolled over into the vineyard was that destiny? Or just shit happening?

I remember a tranquil morning about a thousand days ago when two planes filled with passengers dropped out of the clear blue sky and crashed into each of the Twin Towers, lodging inside the buildings seventy stories in the air, eventually causing them to collapse, vanish from the skyline, and killing thousands of innocent people who were just going about their business. What was that . . . ?

"*E allora?*" prods Ivana's mother, stirring me from my thoughts. "*Il destino? Tu credi?*"

"No," I reply finally. "I don't believe I do."

BY THE BEGINNING of December the transformation is nearly complete. A dark veil is draped over everything. The *nebbia* has taken up residence and sits thick as cotton in the valleys or clings stubbornly to the hills. Sometimes, looking down at the milky fog filling the crevices with only the peaks sticking up above, it seems as if the geologic clock has been turned back eons and that the vine-covered hillsides are once again submerged under water. Other times, the dense mist creates a sort of limbo land in which everything familiar is obscured and neutralized.

The short days are as interminably long as they are dark and dreary and cold. Castiglione seems like a deserted ghost town, shuttered tight, with only the gauntlet of barking dogs to acknowledge the presence of a stray passerby. Even Renza closes up early and goes home, cursing the *tirchiacci* who prefer to stay inside and drink their own wine or coffee rather than venture out and pay to drink hers.

The truffles have dwindled and then disappeared altogether; the *trifulau* forsake the shadowy woods for the comforts of the stove while their sleeping dogs twitch, dreaming of hunts past, of the pungent smells that rose up out of the moist earth beneath their scraping paws, and of the windfall of stale-bread treats that followed. Now the only ones to disturb the forests are the occasional pack of orange-vested, gun-toting hunters in search of wild boar.

The vineyards too are left alone. The grapes have been picked ages ago (the sweaty harvest seems but a distant mirage) and made into wine, which is now in deep gestation; the pressings have been distilled into grappa, a welcome nip of which helps prevent the coldness from getting in. The lethargic vines are thin, dry, brittle, and draped with dead leaves, like an old lady in mourning. The earth is cold and hard, nearly frozen. Le Munie is barren. And the months stretch out ahead like a yawning abyss.

Once again, my internal thoughts mirror the surrounding bleakness: death, past opportunities squandered, mistakes made, alienation, a nameless void, anxiety, and desperation.

But the darkness is not absolute.

In this long, desolate night of winter, two stars twinkle benevolently in the blackened sky. One of them has a Japanese name—Yamaha. I finally managed to get myself a piano; actually, I only rented it and it was only a digital piano (no strings). But it was fine: it had all eighty-eight keys, was easy to get upstairs, and had volume control so I could play it whenever I wanted. Most important, it allowed me to become reacquainted with two of my old friends, Bach and Chopin.

The other star has a name too. It's called Love.

Ivana and I had known each other for just over a year now. The anniversary of my return for New Year's—which signaled that what had occurred between us that first November was not going to be just a short, isolated adventure—was fast approaching. Late one night during that first return visit Ivana had asked, "So, you are in love with me?" She had assumed I came back for her—and she had probably been right (along with my attraction to the vineyards, the desire to escape the dreaded holidays at home, and the lust for physical closeness). But *love*?

Her bluntness took me aback and I stuttered, "Well, I uh, I don't know, I feel . . ."

"Because if you are, it's a problem, no?"

Though tough and *allegra* on the surface, she was still recovering from her broken marriage. Love—the word, the feeling, and the vulnerability it brought with it—*was* a problem and it scared her. Truth be told, it was a problem for me too.

My first "real" love was a long one; seven years, from the age of thirteen to twenty, beginning in childhood, ending in adulthood, and bridging the whole awkward span of adolescence. Its end, inevitable though it may have been, was a rude awakening that coincided with other catastrophes of life and left me reeling and cynical: love *could* end, loss could occur, and there was nothing anyone could do about it. Nothing is permanent. We are all ultimately on our own. Shit happens.

Ever since that time, both the word and the idea it represented were problematic. My first experience set a precedent that was nearly impossible for anything else to live up to, bound up as it was with the loss of childhood innocence and naïveté. We had used the word easily, naturally, almost automatically, like breathing or buttoning up your shirt: "Love ya," "Yeah, love you too," twenty times a day year after year. We just assumed it would always be.

Then one day it wasn't; it was gone, vanished in a puff of smoke. Afterward, the word carried with it the painful recollection of what was not. So I erased it from my vocabulary. To have used it would have felt like a lie.

Not that the feeling was never there. There were others, to be sure; I loved them then—some of them I still do. But I couldn't say it. "Words are only words," I reasoned. "Feelings are another matter. Just because you don't say something doesn't mean you don't feel it." That is certainly true. But sometimes people need to hear something said aloud, need to have it confirmed and reinforced even if they know that it's there. And if you don't name a feeling and proclaim it from time to time, how can you yourself know whether it is there or not?

Enough is enough, I thought. About six months after New Year's, not long after I had moved here, I threw caution to the wind and I said it.

"*Ti amo.*"

For me, using this word was an act of faith, a reckless leap into the unknown, an acknowledgment and an expression of something I felt even though I didn't know exactly what that something was. I knew it didn't mean what it had meant the first time and I knew that saying it didn't necessarily mean that it would last "forever." On the contrary, I knew that it was something delicate and volatile and perishable—and that can be scary. To "name it is to claim it," and claiming something carries with it the possibility that it might be lost or taken away from you, or not reciprocated. I knew all that but I said it anyway.

"*Ti voglio bene,*" she replied.

"*Ti voglio bene*" is a term of endearment and affection, literally "I want you well" (though scratch out the *bene* and it becomes "I want to screw you"). It is used by really good friends for one another, by mothers for their children, and by kids in the early stage of a childish platonic romance. It is a sincere expression of affection. But it is not *amore*.

That was okay. It had taken me long enough to muster up the courage, and I wasn't going to retreat. I didn't say it often and I never said it casually or automatically, but I did say it. And each time I did I could see the consternation come into Ivana's eyes, watch the lurking fear and discomfort drift in like a storm cloud, and practically feel her internal struggle.

"Do you mind when I say that?" I asked her once.

"No, of course not. I like to hear it. *Anch'io, ti voglio bene.*"

One night in mid-December, just after I had gotten back from a trip to New York, we were lying in bed surrounded by the hushed silence of the Langhe winter. While I was away, Ivana had gotten us a small fiber-optic Christmas tree. It was certainly a far cry from the aromatic seven-foot Frasier fir I got from the Canadian guy who appeared on the corner each year right after Thanksgiving, hauled up to the eighteenth floor, and managed to decorate just in the nick of time. But I liked it; it made a comforting little clicking noise as the internal wheels spun round creating ever-changing colors and patterns that danced large upon the darkened walls.

"*Ti amo,*" she whispered.

"What?" I asked.

"I said I love you."

"You do? Are you sure?"

"Yes. I realized it while you were away. I thought about you, I missed you; I worried about whether you were okay. I love you."

The quiet but firm words echoed through the silence of the night.

"I love you too," I replied as the tree purred softly, the colors continued to dance, and we drifted off to sleep in each other's arms.

WE ARE RUDELY woken from our deep winter slumber by the alarm clock going off at the ungodly hour of five A.M. We drag ourselves out from under the toasty covers, feel a shock of icy reality as our feet touch the frigid tile floor and an even nastier shock as we splash cold water on our puffy faces.

Still catatonic, we throw on layers of clothes and heavy coats, get in the car, and set out into the dark, frozen, still-sleeping day. Why not stay cozily cuddled up in bed? The answer is simple: this is the second Thursday before Christmas and we are on our way to Carrù.

I have been here before, but never for the Fiera del Bue Grasso (Fair of the Fat Bull) and was advised to come early. When we arrive at just after six, it is immediately apparent that something is up. This normally sleepy little village is already astir with excitement and bustling with activity. People are busy setting up market stands, smoking, and talking in loud daytime voices, and the smells of fresh hay and manure, mixed with the occasional welcome burst of espresso each time we pass a bar, hang heavy on the frigid air.

We make our way over to the (usually deserted) open-market area to find even more commotion going on. Farmers, trying to unload giant beasts from cramped little trucks and maneuver them down wobbly ramps and into assigned stalls under the big market roof, are pushing and pulling and whacking them with small wooden canes and yelling at them in Piemontese. Eventually, the reluctant, stubborn animals give in and go bouncing merrily down the flimsy gangway. Inside, the rapidly filling pavilion resembles a white sea of undulating bovine flesh. Periodically, the frantic call *"ATENSIUN!"* thunders through the air, and a gap spontaneously opens in the milling crowd as a new addition comes barreling in, thrusting pointed horns, blowing steam from its nostrils, and effortlessly pulling its handlers along behind it.

These are impressive animals indeed, from the "little" 500-pound calves to the snorting bulls. But most impressive of all are the massive *Buoi Grassi* (Fat Oxen); they are the size of a small truck and resemble a white rhinoceros without the horn. Like champion sumo wrestlers, these animals proudly flaunt their rippling corpulence, which is duly admired by the gaping spectators. They are the undisputed stars of this fair, not to mention an essential component of authentic *Bollito misto alla piemontese*.

Having scoped out the scene but not quite adjusted to the morning chill, we decide to duck inside to warm up. At just after eight, the venerable Ristorante Moderno is not only open but nearly full. People seated at long tables are already well into a breakfast of stewed tripe and boiled meat; copious bottles of Dolcetto di Dogliani are already half empty, and a trio of musicians—accordion, clarinet, and tuba—plays popular folk tunes with which many, when not otherwise busy eating or drinking, sing along.

"*Ciao, Alan, come va!*" shouts the flustered chef/owner Bruno over the hubbub. There are bags under his eyes, an indication that he has probably been up all night standing over large bubbling pots in the kitchen. This is his big day too. "*Venite,*" he says breathlessly, "*venite pure.*"

We slide into a couple of empty seats at a crowded table. Without our saying a word, bowls of molten Minestra di Trippa appear before us, and we greedily dig in. The tripe has an earthy but subtle (almost sweet) flavor, and the soft strips of honeycomb melt away in our mouths. The hot soup has a sort of microwave effect, warming from the inside out, and as the heat reaches the outer surface of my skin, not only are the last traces of cold dispelled, but I actually break a sweat and have to peel off my thick sweater. Someone at the table keeps re-filling my wineglass without my noticing it, and as I soak up the last drops of broth with a piece of bread, I am beginning to feel quite content indeed.

Then a cart piled high with assorted cuts of steaming boiled meat rolls out of the kitchen and stops right beside us.

"No, really, I can't," but apparently knife-wielding Bruno doesn't hear me, for in no time big plates of moist, vaporous meat are assembled and placed proudly before us and a tray packed with sauces and condiments shows up as well. As it turns out, I can after all.

We delight in the seemingly infinite combinations of sauce and meat, a veritable symphony of taste and textural possibilities, while my glass of Dolcetto di Dogliani is never empty (despite valiant efforts to make it so) and the band plays on.[1]

At a certain point, neither of us can take another bite, and I finally have to concede victory to the nearly full bottomless glass of wine. We are almost too warm and comfortable and in danger of melting into the chairs. It is time to rouse ourselves and venture back outside.

When we get up to leave, the whole room breaks into a rousing rendition of "New York, New York" (apparently word has leaked out that there is an American in the house). We have a bracing espresso at the bar while we pay the bill, and as we finally stumble out the door many, many more people come stumbling in.

Outside, the sun has finally come out and the chill has vanished. The festival is in full swing now, and the streets are packed with people from all over the Langhe. We slowly make our way back to the covered pavilion where the animals, all lined up as if for some bovine beauty pageant, bellow contentedly while their masters boast about "how big theirs is," and judges work their way up and down the rows, writing their best picks right on the animals' sides with a thick wax marker.

After all the judging has been done, winners in each of the categories are led one by one into a large ring where prizes are awarded by a host of dignitaries. While the rest of the animals squeeze back into their trucks and head for home, the top ten are paraded through town in a procession led by baton twirlers and the Carrù marching band. When the parade is over, the last of the Fat Bulls leave town and the ninety-fifth Fiera del Bue Grasso begins to wind down. But it was *not* over. The huge tent in the main piazza set up by the *comune* to dispense *Bollito* to the masses is full to capacity and has long lines of hungry people outside waiting to get in, every restaurant is packed to the gills, and inside Moderno, carts piled high with boiled meat continue to roll, wineglasses are never empty, and the band plays on and on.

· RECIPE #24 ·

MINESTRA DI TRIPPA

Tripe Soup, adapted from Ristorante Moderno, Carrù

TRIPE SOUP FOR breakfast? Yes, or for lunch or dinner. This wonderful cold-weather dish makes its appearance each fall and stays around until the snow melts. One of my favorite occasions to eat it is a festival in Dogliani, the Fiera dei Santi on All Saints' Day, where a chickpea-filled version prepared by villagers is ladled out of big pots into individual bowls (*scodelle*) to throngs of half-frozen people, who consume the steaming contents at tables under the open-roofed market from eight A.M. until noon.

Tripe is the honeycombed lining of a cow's (or pig's) stomach. Not a very appetizing thought, to be sure, and for that reason most Americans shy away from it. But slaughtering animals and consuming their flesh is not that appetizing either, if you think about it. Luckily most of us don't have to. Tripe is available at most good butcher shops already cleaned and parcooked, so it is easy to work with. More important, the flavor is unique. It is more delicate than most other offal, and properly cooked, it melts in your mouth—another primarily textural experience.

Following is the recipe from Ristorante Moderno, where tripe is almost always available, along with just about every other part of the animal!

INGREDIENTS

3 tablespoons extra-virgin olive
  oil
1 piece of pork belly or fatback
  (about 4 ounces), rind
  removed and cut into ½-inch
  squares
2½ pounds tripe (beef),
  preferably fresh, cut into
  strips ¼ inch wide and 2 to 3
  inches long
4 leeks (see note below),
  tough parts removed, split,
  thoroughly washed, and cut
  crosswise into ⅛-inch slices
  (about 1 cup)
4 medium-sized potatoes,
  peeled and cut into 1-inch
  dice (about 1½ cups)

1 small squash (such as
  butternut or acorn) peeled,
  seeded, and cut into 1-inch
  pieces (about 1½ cups)
1½ to 2 quarts veal or chicken
  stock, fresh or canned
Coarse salt
4 to 6 tomatoes, peeled, seeded,
  and diced (about 1½ cups) or
  substitute 1 12-ounce can of
  diced San Marzano–type
  tomatoes
Black pepper, and crushed red
  pepper to taste
2 tablespoons fresh parsley,
  coarsely chopped (optional)
Grated Parmigiano Reggiano or
  Grana Padana cheese

1. Heat the oil and pork belly in a large, heavy-bottomed pot on medium heat. Add the tripe, and cook gently for about 10 minutes. Stir in the leeks and sweat until they just begin to soften.

2. Add the potatoes, the squash, and the stock (there should be enough liquid to cover the ingredients in the pot; if not, add additional stock or water to cover).

3. Add 2 teaspoons of salt and bring to a boil, skimming off any scum that rises to the surface. When it reaches a boil, lower heat to a simmer and cook, stirring occasionally, until the tripe is cooked (approximately 2 hours). To test for doneness, taste a piece of the tripe; it should be soft and succulent; if it is tough or rubbery, continue cooking. By this point, the potatoes and squash should have broken down somewhat and thickened the soup; if not, smash them up a bit with a wooden spoon or whisk.

4. Add the diced tomatoes and cook for another 10 minutes. When ready, the soup should be moderately thick and the tripe melt-in-the-mouth tender, but firm and resilient (not mushy) and still holding its shape.

5. If it is too thick, add more stock. And add more salt if necessary, black pepper, and crushed red pepper to taste. Stir in the parsley (optional) and serve in individual bowls with a drizzle of olive oil on top and grated cheese.

YIELD: SERVES 6 TO 8 AS AN APPETIZER OR 4 AS A HEARTY BREAKFAST.

NOTES: This soup may be made several days ahead in large batches and reheated as necessary. It may also be separated into small containers and frozen.

At Moderno, Bruno uses leeks from the nearby town of Cervere, which is famous for them. Each spring Cervere hosts a festival celebrating these leeks. And they are gorgeous indeed: incredibly long (3 feet is not uncommon!), alabaster white, and unusually tender and sweet, they are proudly offered for sale by farmers in thick bunches tied up with string.

40 · NIGHT OF THE WISE KINGS

"IT LOOKS LIKE we're going to have a real winter this year," people say. "A winter like there used to be twenty years ago, before the world started to warm up."

Precipitation—rain, snow, rain and snow together—falls from the sky and moistens the still-parched earth. It snows a couple times even before the end of November and in the first week of December there is a storm that sticks. It is unrelentingly cold and damp; it seems as if our summer wishes for a little cool relief have been belatedly answered tenfold.

In the midst of this numbing chill, a third star appeared, shining brightly in the otherwise long, dark winter night. Next to my apartment, over the road coming into Castiglione, hangs a huge red shooting star with the words BUON NATALE! in its tail, just like the one that guided the Three Wise Kings to the little village of Bethlehem. Colorful neon decorations—a candy cane, holly leaf, snowman, jingle bells—adorn the main street, everyone has been saying *auguri* for weeks now, and village mutts howl in harmony as if rehearsing carols for a canine glee club.

An expectant hush falls over the village as the holiday nears.

On the frightfully frigid night of the twenty-third, Ivana and I venture out to see the *Presepio Vivente,* the living crèche.

"Okay, we can go if you want to (although if you would prefer to get a movie and stay home, that's okay with me too), but we have to get there early; it will be a real *casino* later on, and I don't want to have to park far away and walk in the freezing cold in order to get there."

When we reach Dogliani just after dark, we are met by an army of *Protezione Civile* in orange vests and fluorescent yellow pants waving us on with lighted batons up the twisting road to the upper (old) part of town and into a nearby makeshift parking lot. We are the very first in line at the entrance but have to wait outside anyway, as it doesn't open for another half hour. On the other side of the gate, every obvious sign of the modern world—electric circuits, mailboxes, ATMs—has been covered up, and over the entrance way is a hand-painted sign decorated with shooting stars: BETLEMME.

At just past eight thirty, torches sticking out of the walls are lit and the mayor of Dogliani comes on a loudspeaker welcoming everyone to the *Presepio Vivente.* Electric power is then cut off, and we are allowed in. The town has been transformed into an Old Testament village (or at least someone's interpretation of one). We stroll along narrow cobblestone streets lit by flickering flame and peer into garages and cantinas that have been converted into Old World workshops. There is a candle maker, a scribe, a fortune teller, and a baker; a blacksmith, a wood-carver, someone preparing potions of herbs and spices, and someone else making barrels. There are people roasting chestnuts, which they hand out to visitors in small paper bags, and miniature bonfires burning in low iron braziers along the streets. Everyone is dressed up in either medieval or Middle Eastern costume.

"Let's go find the *panetteria,* I'm starving!" Ivana says, and sure enough, plopped down in the middle of a courtyard is a wood-burning oven filled with tiny flatbreads, which a large man in tights extracts and baptizes with oil before passing them out to the eagerly waiting crowd. We practically burn our hands on the hot bread, and gusts of steam spew from our mouths as we devour it. Nearby, someone in a turban is stirring polenta and someone else offers *crostini* of stinky *brus* cheese, both of which we devour as well. Hunger slaked, we continue our tour of Bethlehem, passing a basket weaver and a vine-grafter, and walk by an old-fashioned tavern (*osto*) where farmer-types in dusty work clothes and "Egyptians" in caftans are drinking wine, smoking, and playing cards.

As we move on, I can't help questioning this weird jumble of time and place: just what kind of Bethlehem is this supposed to be? Ancient Egypt meets archaic Piemonte? Don't these people realize that there were no *osto*s at the time and place of Christ's birth? But this childlike naïveté, this temporary suspension of belief in the logical order of things, is precisely what makes the *Presepio Vivente* so enchanting.

At around ten, a blond "Mary" (who seems to have had her hair bobbed and highlighted in some not-too-ancient salon just that morning) turns up, rosy-cheeked and tummy bulging, astride a donkey led by "Joseph." They go from stand to stand, asking if anyone knows of a place where they can spend the night. No one does; everything is full. Later, they find a place in a makeshift hay-strewn stable filled with animals, adjacent to the old castle. And there around midnight, a real baby will appear cradled in Mary's arms for all to see.

And why not? In this rural area the very idea of the nativity seems to make more sense than it did back in New York City. Up until recently, almost everyone here kept farm animals, and many remember spending cold winter nights in the stables where soft hay and animal warmth provided more comfort than drafty farmhouses. People here *know* firsthand what a manger is. And it takes only a small leap of the imagination to envision a couple of travelers (Macedonian immigrants perhaps?), one of them a woman far along in pregnancy, caught without a place to sleep on a cold, starry winter night just like this one, taking refuge in a warm, smelly stable. Right here in Dogliani.

When the manger is open, a large star high above it is lit to signal the birth. Do three Wise Kings show up as well? I suspect they do, but I don't know for sure because we don't hang around. By eleven the Madonna-and-Baby Jesus line is already quite long, and Ivana is frozen to the bone. Plus we are hungry again (getting a decent parking place meant missing dinner), and all the nearby restaurants are either full or closed; there is no room at the inn for this pair of cold, hungry travelers. So we get in the car and head back down to our local pizzeria, where, as luck would have it, we are welcomed warmly. The heat is steam, the bright light is electric, the music canned, and there is not a caftan in sight. The oven, however, is fired by wood (just as in the olden days), and the hot cheesy pizza—washed down with a bottle of Dolcetto di Dogliani—is just fine.

The next night, Christmas Eve, we go to Midnight Mass at the church in Monforte, whose name, Madonna of the Snow, now makes more sense.

Inside, the church seems terribly large (which it is, compared to the lovely little church of Castiglione) and almost garish in the harsh lights. And it is packed.

The priest is new. He has already offended a large portion of the parish by deciding to start the Midnight Mass an hour earlier than usual, and you can feel the bad vibe in the air: "Just who does this young upstart from God-knows-where think he is anyway?" you can practically hear the stone-faced matrons of Monforte saying. Before long, he offends me too by giving a wordy, uninspired sermon—on Christmas Eve of all nights! He mumbles through the litany as if emitting a bored sigh; he holds up the Host and the Chalice—the bread of heaven and the cup of salvation—nonchalantly as if performing some sort of routine calisthenics, and the parishioners respond by falling into line robotically, like sheep being herded into a pen or kids unenthusiastically lining up to take their medicine.

I fall in too. The thin, tasteless little wafer sticks to the roof of my mouth like a spongy blister and there's not even anything to wash it down with. Bread of heaven indeed, I think to myself.

The informal "communion" that takes place after the service out on the icy piazza is better. There the local branch of the Alpini offers toasted *panettone* and Vin Brûlé to one and all, and everybody crowds into the little square wishing one another *auguri* and stamping their feet to keep them from freezing.

Afterward, with the hot spicy wine melting a channel to the middle of our frozen bodies, we hurry home to Castiglione, jump into bed under the heavy down cover, and squeeze together, dispelling the last traces of chill, as the lights of the fiber-optic Christmas tree dance and sputter like sugar plums through the night.

· RECIPE #25 ·
VIN BRÛLÉ

Hot Spiced Wine

INGREDIENTS

¼ cup sugar
2 cinnamon sticks
4 cloves star anise
2 cloves

6 white peppercorns
2 dried bay leaves
4 glasses (3 cups) red wine

1. Put the sugar in a saucepan along with cinnamon sticks, star anise, cloves, peppercorns, and bay leaves. Add the wine, stir to mix, and turn the heat to medium.

2. Let the wine gradually heat up, stirring occasionally to dissolve the sugar, but do not let it come to a boil.

3. When the wine almost reaches the boiling point, lower the heat and let it steep for 15 to 20 minutes, making sure to inhale the perfume of the spiced wine as often as possible.

4. Discard the bay leaves and other spices before ladling into mugs.

Serve piping hot with cornmeal biscuits *(biscotti di meliga)* or toasted
panettone.

YIELD: SERVES 4 TO 6.

––––––––––––

NOTE: You can use pretty much any wine you like for this, however a
young, medium-bodied wine with moderate acidity and soft tannins
such as a Nebbiolo d'Alba or Barbera work best.

41 · INFUSIONS

VIN BRÛLÉ IS made by placing herbs and spices into red wine, heating
it up and then letting it steep, infusing the flavors into the wine. This
marriage of wine and spice is prepared often during the winter months
and is an excellent antidote to the cold weather. Whether it is simply
the promise of ingesting hot liquid, the naturally curative properties of
alcohol (some, but not all of which has evaporated into the surrounding air) or the invigorating aroma of spices (called *droghe* in Italian), a
huge pot of it steaming over a wood fire is practically guaranteed to
chase away the chills.

There are many such alcohol-based concoctions in this grape-
growing, herb-rich sub-Alpine region, and almost all of them have
long been reputed to have positive medicinal properties.[1]

Up until not too long ago the distinctions between doctor, pharmacist, winemaker, and concoctor of medicinal potions were blurred
(sometimes the very same person held several of these titles); every
pharmacy stocked a variety of alcoholic concoctions behind their coun-
ters, and their use was often prescribed by doctors. Barolo has its own
unique version, ubiquitous throughout the zone but practically unknown outside it, called Barolo Chinato, a fortified *digestivo* officially
classified as an "aromatized" wine.[2]

Like many things around here—truffles, taxes, recipes—there is great secrecy and mystique surrounding Barolo Chinato. The basic ingredients are known by all. Barolo wine, grain alcohol, and quinine[3]—*china* (pronounced KEY-na) in Italian—which gives the concoction its name. But the numerous other ingredients (in some cases as many as twenty), their amounts, and the exact process employed in marrying them all together—is thoroughly proprietary and strictly top secret.

Chinato's production is tightly regulated by law, and this is understandable. In an area like Barolo in which a long list of regulations are strictly enforced to protect the appellation, the very idea of adding or infusing foreign substances to a classified wine is viewed as a potential threat. After all, Barolo Chinato carries the name Barolo and thus the entire *denominazione* along with it. Just having the ingredients in a winery where Barolo is made could be considered suspect; there is always talk of doctoring wine or adding stuff to bulk it up during poor vintages, a practice strictly against the law.

Many winemakers made Chinato on the sly for their own consumption—some still do. One of the top Monforte wine producers recently gave me a small label-less bottle wrapped in plain paper. "Here. Take this," he whispered. "I think you will enjoy it. It's my grandmother's recipe. But don't open it until you get home and don't tell anyone where you got it. They could shut us down."

Ivana's father, Francesco, made Chinato. Tiziana remembers him stirring the warm infusion in a big vat and how his sister Catterina used to go to a shop in Torino to buy the secret ingredients. One evening, I actually got to taste it.

After supper Fabrizio disappeared into the cantina for a long time. When he finally emerged, he was carrying a big dusty magnum with a rubber-rimmed metal stopper that he cleaned off, popped open, and tipped into our empty wineglasses. "Barolo Chinato," Fabrizio announced. "My father made it, but I have no idea when." What came out was dark orange bordering on brown, much of the fruit was gone, and it tasted a bit thin and medicinal, but was deliciously beguiling all the same. It suggested an old faded wine still struggling valiantly to be bright and jovial along with a spark of pepperiness, a bright hint of orange rind, the wisdom of bitter almonds, and a kind of rubbery astringency. It was unusual but lovely; it cut right through our full stomachs and, despite its apparent feebleness, seemed to effortlessly tamp down our big meal.

You need a separate license to make Barolo Chinato, and few actually have one. For most winemakers it doesn't make sense to incur the extra expense in order to produce such an obscure product. One of the original licenses belongs to the historic Cantina Borgogno in Barolo. The *azienda* is run by two brothers, Giorgio and Cesare, who alone are personally responsible for making the Chinato; one of them knows the secret recipe and the other knows the secret procedure. Their individually classified information will be passed on to their respective children.

The most legendary producer of Barolo Chinato, Zabaldan of Monforte, was not so fortunate. Though he died over a decade ago, people still talk reverentially of his exceptional Chinato: "Zabaldan?" one person told me. "Certainly I knew him. He was crazy but his Chinato . . . there was nothing like it and never will be." There are purportedly still some bottles floating around but I have not been able to find one and will probably never get to taste it: Zalbadan had no sons, and as he would not entrust his recipe to anyone else, it died with him.

Of the handful of people currently making Barolo Chinato, the best is Cappellano. The *azienda* is located in an undistinguished modern-looking house in the Commune of Serralunga right across the road from the entrance to the historic Fontanafredda estate. When I called to make an appointment, a young man answered, "Ah, yes. It is probably better that you should speak with my father," and gave me his cell phone number. "Yes, yes, yes," said a flustered but not unfriendly voice when I finally got through. "I am very busy now, there is always so much to do, but yes, I think I can make some time for you." We set a date.

The present Cappellano, Teobaldo, is the great-grandson of the very man credited with inventing Barolo Chinato. Like all his male forebears, Teobaldo was trained in medicine and his name carries the distinguished title of *dottore,* though these days his practice is pretty much limited to tending vines and looking after his cantina, where they make wine as well as Chinato.

On the appointed day at the appointed time I pulled up in front of the house, rang the bell, and announced myself over the intercom. A moment later, the big gate magically swung open and I rolled slowly into the courtyard and parked.

At the front door I was met by a tall man in thick corduroy pants and a pale orange sweater over a plaid button-down shirt. Eyeglasses hung on a string around his neck and a receded hairline displayed a huge, wrinkled forehead. His head tilted invitingly and he looked down at me with a big benevolent smile mitigated by eyes that seemed somehow sad and melancholy. "Hello. I am Cappellano, Dr. Cappellano. Won't you please come in?" he said as he stepped back and held out a long arm, ushering me into his study.

The room was large, square, and wood-paneled. There was a big desk in the center with a small bronze sculpture on it and several upholstered chairs in front, one of which I settled into. The walls were covered with paintings and certificates; there were lots of books, a shelf displaying bottles of wine, and a large birdcage in one corner where a couple of parrots twittered away. The partially drawn blinds in front permitted broken slats of afternoon light to filter in.

He poured us each a small glass of Chinato. I had tasted it before, so the wine was familiar to me (Chinato, like most medicines, fortified wines, or spirits where a set recipe is adhered to, doesn't change a whole lot from one batch to another): it was remarkably balanced and complex. It was Barolo, yes, with the body and fruit and tannins characteristic of the wine, but this drink had a whole host of other flavors as well; pepper and nutmeg, clove and allspice, star anise, orange rind, licorice, coffee, rosemary, mint, and cherry cough drops. Most amazing of all, this long roster of flavors was present in perfect order and harmony— no one stood out from or overwhelmed the other.

I knew better than to even ask exactly what was in it. We drank nonchalantly, hardly mentioning the wine before us, and we talked. Teobaldo is a humble, self-effacing giant; he talks slowly and quietly— his words were just audible above the twittering of the birds—but he expects to be listened to and likes to guide the conversation.

We talked about the international wine market, the "Americanization" of Barolo, and the current economic crisis in the States. We talked about the ongoing project of turning the two hills into one (something that disgusts him), the relentless "upgrading" of the old vineyards, the steady conversion of ever more land to vine and the problems it was creating. And we also talked about literature and music and painting, things about which Teobaldo is passionate.

I tried to subtly shift the conversation toward what had brought me there, Barolo Chinato. And finally he yielded a bit.

"Yes, well. My great-grandfather was a doctor and a pharmacist. At that time wine and spirits were closely associated with well-being, just like a medicine is, and the positive benefits of certain herbs were well known. We had some vineyards here in Serralunga and my grandfather had some patients with chronic indigestion, so he put the two together and came up with Barolo Chinato. The recipe for it has been passed down through my family for over a hundred years. . . . Do you like this painting here?"

Not so easily deterred, I asked about Zabaldan and lamented the fact that his Chinato has disappeared forever.

"Yes, it was quite good indeed, and quite different from ours stylistically. But I completely understand and respect his reasoning: Zabaldan made lovely Chinato; Zabaldan is no longer here and so it is right that his Chinato shouldn't be either. It is like the Zen monks: you write a poem, place it on a river, and it goes drifting down and away forever. Zabaldan *was* his Chinato, just as I am mine. I am happy to have a son I can pass my recipe on to—he, you might say, is my river—but if I didn't, I would have done exactly the same thing."

There was a brief pause, then:

"Please," Teobaldo said courteously, rising up from his seat like a polite physician who has finished his exam and has a room full of other patients waiting. "Please excuse me, but I really must get back to work now. I have enjoyed speaking with you. Do feel free to stop by anytime and we'll chat some more. But let's leave the topic of Barolo Chinato alone, shall we? On that subject I believe we have said just about all that there is to say."

AFTER NEW YEAR'S comes Epiphany (January 6) and with it comes *La Befana,* a combination Santa Claus, scary witch, and benevolent grandmother. Old and ugly, hunch-backed and hooked-nosed, she travels on a broom bearing gifts for the good children and coal for the bad. Shop windows, shorn of Christmas items, are filled with *Befana* dolls and bakeries sell colored confections called *carbone dolce* ("sweet coal"). After Epiphany the holidays are definitely over: all the decorations come abruptly and unceremoniously down (I, however, resolve to keep the fiber-optic Christmas tree up at least until the snow melts), useless gifts are stored away, leftover *panettone* is discarded, and life returns to usual. I, along with everyone else, go back to work; I return to the vineyards.

In mid-January we begin pruning and it seems to go on forever.

It has snowed a lot this year. There have been several major storms, stranding traffic, closing schools, and forcing people to remain indoors. Many evenings, after supper at home in the Bussia, Ivana walks to Castiglione in the thickly falling snow rather than risk driving up the steep, slippery, unplowed road. And when she does, I come to meet her halfway in the deserted snowy stillness, our figures eerily emerging into view through the luminous mist.

At Cantina Parusso, when it's snowing we usually stay inside cleaning or fold-and-stapling cardboard boxes in anticipation of some future order. But on days when it's not, we venture out into the icy grayness.

We go tromping through the dormant vineyards, our heavy footsteps echoing one another and creating dirty paths in the otherwise pristine snow. It is *cold.* I wear layers: three pairs of socks; T-shirt, turtleneck, sweater, down vest and jacket; hat and gloves. But it is *still* cold. You can't feel your fingers until around ten A.M. when, thanks to continuous motion and circulating blood, you begin to feel an itchy throbbing. Feet, however, are hopeless; either submerged in the snow or slip-sliding on the ice, they never warm up.

You pray for just a bit of sun, and if it comes, soak it up greedily, like a dry sponge tossed into a warm bath. Other times, when the sun doesn't come, you pray for a blizzard so you can go home or at least go into the cantina where it's warmer. Often the day passes without either, going from icy darkness to cold light-filtered gray to icy darkness again.

On these days, the only respite, the only thing that gets you through the day, is lunch. It lasts an hour and a half. At just after noon, we trudge back to the cantina, remove icy boots and the outermost layers of clothing, slip on a pair of spare shoes, and go into the house, into the kitchen, where it is warm and steamy. It takes a while to thaw out, but thaw out we do. There is always plenty of food—a big bowl of pasta or a hot soup packed with melting meat and vegetables, followed by a roast or a stew or a piece of *bollito*—lots of wine and then coffee, while we watch the news or some silly game show on TV. Sometimes Rita prepares a thermos of sweet, hot coffee to take with us.

Then it's back out into the cold.

Francesco and Enrico are well accessorized with different boots for different conditions, including those big puffy Moon boots for walking through snowdrifts. I settle for a pair of tall rubber boots I pick up in Monforte, which, along with my three pairs of socks, helps keep some of the dampness out. I also get myself what is called a *"Tony,"* the classic blue worker's jumpsuit, to go over my layers of clothing (and under my down jacket) and a wool cap to cover my ears. Then I go to work.

This pruning is different than it was last year with Fabrizio in Le Munie, and not just because it's colder this winter. This is not just a pleasant, novel pastime; this is a job. Day after day I go trudging out into the cold dark and hour after hour I yank cut branches down from the wires until it's time to go home again and thaw out. The basic activity—grab and pull—is familiar: I did it last year when I pruned Le Munie, the first time I stepped foot into the vineyard and confronted the vines, and again this past fall when we stripped the green vines bare with Bastian. But everything else is different, as if I just woke up from a warm pleasant dream and found myself out here in the freezing cold.

I grab a bunch of dry, brittle branches in my numb gloved hands and pull. Eventually they give way, I lay them down in the center of the frozen aisle, and move on to the next vine and the next, vineyard after vineyard, day after day. There are twenty hectares to do instead of just one and here I don't get to make the crucial choices of selecting the *capo a frutto* and the *portao*; I don't get to help plan and shape the future of the vine. I just grab and pull.

This is as it should be. Francesco and Enrico are both much more knowledgeable, experienced pruners than I am (Enrico, at twenty-eight, has spent half his life here at Parusso; his father worked for Marco's father). They know these vines, their history and particularities; they know where they are at in their life cycle and how best to take care of them, for they have done this many, many times before. I watch what they do while I'm yanking down the cut vines in their wake, analyze their choices, and learn from them. Sometimes I ask questions, but mostly I try to figure out the answers on my own.

The starkness of winter in the vineyards can be lovely and strange. We are usually the only ones around and when we do see someone else, we salute him or her like fellow survivors floating on a vast white sea after a storm. Sounds seem to bounce off the layer of snow and be absorbed by it too, sounds like the church bells of the Bussia (recollections of the long-ago summer feast) or lonely bird calls. I like to watch the gradual progress of our work transform the cluttered fields of dried-out vines into clean lines of single squiggles standing out against the whitened background. When it's not too cold and the wind is not blowing too hard, it can be quite nice to be out here.

But it is also tiring and tedious, both the work and the season. Even Signora Rita gets fed up. Mid-winter, it is decided that she won't cook lunch for us anymore. We three employees significantly increase the amount of work required to put lunch on the table (not to mention the additional expense), and without us, they could do something much simpler. They had been talking about making this change for two years; the winter doldrums just tipped the balance.

So the three of us start going to Monforte for lunch. Terra e Luna (formerly Al Ponte, which is how many still refer to it) offers what is called a *pranzo di lavoro,* a worker's lunch: antipasto, primo, secondo, cheese, fruit, dessert, wine, water, and coffee for 10 euros per person. It's cheaper than eating at home; I can fill up at lunch, have a light supper at home later, and actually end up saving money. Plus the food is good.

Italy is a Catholic country; nearly everyone is automatically baptized at birth and crucifixes are everywhere (a Muslim family in Rome recently sued to have them removed from a public school, creating a national uproar—they lost). Only a moderate number of people attend church regularly but that's okay; God understands. Everyone, however, has lunch. Lunch is sacred. Those who can't get home make arrangements for someplace to go, preferably the same place each day (if you can't be with your family, it is at least nice to go somewhere where you are known and expected), but go they do.

Back in New York I often didn't eat lunch at all (I was usually too busy cooking it), and if I did, it was likely to be a lukewarm pizza or soggy salad at my restaurant or a quick bite on the run. Here, however, that is not acceptable. Offices and shops close; *everything* closes—except, of course, places to eat. Eating lunch is looked upon as a basic part—almost a responsibility—of existing in a civilized society, and if you don't participate, you are considered something of an outcast or a renegade.

I didn't want to be either, if I could help it. Besides, I *like* lunch.

Each day when we arrive at twelve forty-five, our table is waiting for us. The place is always packed with pretty much the same people, almost all of them workers in their assorted uniforms (electricians, telephone installers, plumbers, salesmen), and almost all men. The few stray females are looked upon with interest and curiosity, like members of a different species, and are almost always part of a well-dressed group ordering off the menu and drinking wine out of big fancy glasses.

When we sit down, a bottle of wine—usually dolcetto—is placed before us and we drink it out of short tumblers, which double as water glasses. And we never see a menu; the food just comes. It is always simple, standard stuff, and it is always very good. We hang on to the same bowl for the entire meal right up to dessert, cleaning it out between courses with a piece of bread (an action called *fare la scarpetta*, or doing the shoe). The food is served from a large platter by a Romanian waitress with whom we flirt good-naturedly, and if we ever want more of anything, she gladly runs back to the kitchen and returns with another platter.

Usually, however, we don't *need* more of anything; by the end of the meal we're quite satisfied and glad to pass our 10 euros across the bar while we down an espresso, almost ready to head back to the cantina, suit up, and venture back outdoors.

These *pranzi di lavoro* are great; I love the food, the ambience, and the feeling of being part of a fraternity of workers. But they cause me a bit of concern as well. Not only do I *like* lunch, I actually look forward to it as the high point of my day, and the very thought of it sparkles before and after like a sunny oasis in an otherwise frozen desert of drudgery.

Fraternity indeed. In my rubber boots and Tony and wool cap pulled low over my ears I blend in quite well with all the other workers; in fact, I *am* one of them. Sometimes I feel kind of like a conscript doing hard labor in some agrarian foreign service. I remind myself that this is a *voluntary* conscription, that I am doing this because I want to, and I focus on the things I like about it. But basically, this is just a job; I do it to pass the time and help support my being here. Though I've worked since I was sixteen years old, I never felt like I had a "job," like I was working just for money, counting the hours until lunch and saying TGIF when the end of the week finally rolled around. Even when I struggled to get through long hard days at the restaurant, I never thought of it as a job. And now look at me, sweating day after day in the freezing cold for next to nothing until it is time to go home and defrost. What have I become?

43 · *CLANDESTINO/CITTADINO*

A *CLANDESTINO*. WHATEVER MY own personal ambiguities might be, as far as the Italian government was concerned my status was clear: I was an illegal alien. My *permesso*, valid for three months, had expired. After that, the guy-with-friends-at-the-*questura* got me another one, but that was it. Apparently, you can't be a perpetual tourist; you're supposed to go home for six months before coming back and requesting another *permesso*.

I didn't *feel* like a tourist, and I didn't want to go home. So I went "underground." I had my American passport with which I freely came and went (no one at the airport seemed concerned about any permit). But each time I reentered Italy and the eight-day limit passed, I technically became illegal once again.

Throughout the past year, I had continued trying to track down the documents necessary to prove my right to Italian citizenship and, despite some major setbacks, I finally got everything I needed.

I had heard a rumor that, as a resident of Castiglione, I could submit my application here instead of having to go back and do it at the consulate in New York, and (best of all) that this would get me a special nontourist *permesso* that doesn't expire. So one day, I gathered up my dossier and went to the *comune* to find out if this was true.

"Yes, yes. Looks good," said the woman behind the counter. (Apparently, while no American had ever sought Italian citizenship at this office, a number of Argentinean descendants of Piemontese emigrants had, so she was familiar with the process.) "Now you just need to get them translated and have the whole thing validated by the tribunal in Alba."

I found a court-authorized translator who rendered the American documents into Italian. Then I brought the whole thing to the tribunal, where the proper tax stamps were affixed, the pages were stapled together, and a continuous red validation line was applied so that nothing could be altered. I brought this back once again to the *municipio* office.

Finally, everything was in order. They accepted my packet and gave me in return a letter on a Castiglione letterhead stating that I was a resident of the town, that they were in possession of all the necessary documents, and that the process to have me officially recognized as an Italian citizen had begun. This letter was then christened with the town seal, the proper tax stamp was applied, and it was signed by the mayor.

It was now time to come in out of the cold and get a *permesso*. I was not alone. Letter in hand, I crammed into a beat-up Punto with a bunch of other semilegal "aliens" and headed for the *questura*. Our good shepherd was a Moroccan named Salah who had much experience in immigration matters and ran a dingy foreign services agency in Bra. I waited for hours while the other more straightforward cases were dealt with and dispatched. Then my turn finally came. Salah explained the situation and slid my letter from the mayor under the slot along with my application, passport, and expired *permesso*. The policewoman stared at the papers for a while, showed them to her colleague at the window next to her, and a long discussion took place behind the thick glass. She turned back to us and said, "We don't know what to do with this. I have to ask my boss," gesturing to a guy at a desk hunched over behind a tall stack of papers and yelling into a cell phone (I recognized him as the harried man Ivana and I had first spoken with months before). "He's busy now. You have to wait. Next!"

We waited. And we waited.

After a while, the policewoman got up and went over to the desk with my papers. The harried boss looked through them and an animated discussion ensued. Then the two of them went up a flight of stairs at the back of the office.

"Oh my!" said Salah with apparent delight. "Now they're going up to talk to the BIG boss. I knew this was an unusual case, but, as you see, they really don't know *what* to do. It's a gray area (it's not something that happens that much, you know, an American trying to become an Italian). If you're a tourist, you can't really be a resident, but if you're not a resident, you can't apply for citizenship here in Italy. How can you be a resident without a *permesso*? Yet here is a letter from the mayor saying that you are indeed a resident of his town and that you are about to be declared a citizen by the Italian government. Your tourist *permesso* has expired; so what? They wouldn't give you another one, and you're not really a tourist anyway. But is this a renewal or a new thing altogether? It's not clear."

Time passes. The line behind us has come to a standstill. I wait anxiously, frustrated and pessimistic, off to the side.

"No one knows what to do," my advocate says again. "They're probably up there calling Rome for guidance."

"Oh that's great," I say. "Everyone's probably still at lunch. Maybe they should try calling the Pope; he's always on duty!"

Finally the policewoman came back down the stairs and up to the window.

"Come back here please. I need to speak with you."

"Courage!" whispers my Moroccan friend. *"In bocca al lupo!"*

"Crepi," I reply unenthusiastically as I go once again through the door into the inner sanctum, half expecting to be handcuffed, incarcerated, and put on the next plane for America.

"You know," said the young and not unattractive *poliziotta*, once I got back there, "what you have done here is very unusual. You must follow the Italian laws when you are here in Italy."

"Yes, *signora,* I realize that and can assure you that it is my intention to abide by all the Italian laws and regulations. It's just difficult to do so if you don't know what they are. Everyone says something different."

"Yes, well, you must follow the laws anyway. Being here in Italy without a *permesso* is against the law. We could make you leave and put a big X on your passport, and you couldn't come back for another ten years. You don't want that, do you?"

"No, I certainly don't. I want to follow all the laws, really I do."

"What do you intend to do in Italy anyway?"

"Nothing special. I just want to be here. With my girlfriend. She is Italian."

"Ah, I see. Well, just make sure you follow all the laws."

"Yes, *signora*, I will, each and every one. I promise!"

With that I was dismissed back through the door to the other side. After a few more minutes of paper shuffling and a couple big bangs of the rubber stamp, a slip of paper was slid back under the glass. It was green this time and on the bottom were the magic words *"In Attesa della Cittadinanza"*—Awaiting Citizenship.

On the way back, our odd immigrant caravan—two Muslim women, a North African, an Albanian, and an American Italian-in-Waiting, led by a Moroccan who had set up shop in the Slow Food citadel of Bra—decided to stop at McDonald's for a snack. I passed on the food but thoroughly enjoyed the irony of the situation.

With those lightning strikes of the rubber stamp, I went from being a clandestine to a citizen (at least *in attesa*). I was now legally allowed to be here and could pretty much do what I wanted (following all the laws, of course), just like any Italian. I got an official identity card and what's called a *Libretto di Sanità*, which entitled me to basic health services, incredibly cheap medicine, and a doctor in Barolo (the very same one who looks after Bartolo Mascarello). I could even get a real job if I wanted to.

In a sense I had made it: I could remain in Italy indefinitely and was well on my way to becoming a citizen. My existential situation, however, remained pretty much the same. Instead of feeling part of two worlds, I felt part of neither, as if I were stuck in some isolated crack in between and couldn't get out. So I continued yanking down dead vine branches in the icy vineyards while the cold, dark winter stretched on into the future with no end in sight.

PART VI

FLOWERING
AGAIN

44 · RENAISSANCE

ONCE WE FINALLY finished pruning, we moved right into bending and tying the branches, which took place mostly atop a stubbornly lingering layer of snow. There were signs the season might be beginning to change—a whole day of sun that melted some of the snow, creating rivulets through the vineyards, a certain limpidity in the air, the increasing frequency of days when it was warm enough to go outdoors with only two layers on instead of three or four—but it was becoming clear that my internal situation was not.

Though everything around me was in a constant state of transformation, it seemed as if I had hit a solid wall, as if I were caught in a trap and couldn't get out. It didn't make it any easier that the "trap" I was in happened to be in the bucolic area of Barolo, where I had an apartment (albeit a sparsely furnished one), a woman whom I loved (and who loved me), and official residency in a pretty little village filled with wacky characters who now considered me their *paesano*. I was even well on my way to becoming a dual citizen. I had occasional work in the vineyards and cantinas ("How cool!" most people back in the States would say) and even a piano (if only a stringless one). Many would envy me; hey, *I* would envy me!

And yet I wasn't content: something was missing. What had begun as a smoldering postharvest depression had blossomed into some sort of midlife crisis that was on its way to becoming a total identity crisis; what was supposed to have been a temporary transitional phase after the restaurant closed seemed to be turning into a permanent way of life with no end in sight.

Then something happened. It didn't seem like anything at first, just one of those little rhetorical questions you pose to yourself throughout the day. This one popped up one Sunday morning during Mass in

Castiglione. As the priest held up the cup of wine (kneeling altar boy rings bell) and then the Communion wafers (bell rings again), I remembered Christmas Eve in Monforte and thought to myself:

"Hey, what is really going on here? The wine in that chalice is undoubtedly inferior (it's certainly not a DOC wine from the Barolo area), and those insipid wafers are surely mass-produced in some factory somewhere. How odd that these mediocre things should symbolize body and blood. Body and blood of anything, much less of someone some people consider the Son of God made man."

I usually sit on a wooden bench at the rear of the church with the other solo guys, mostly old widowers, and I looked past the backs of the entire solemn congregation to the priest facing us with his arms stretched up in the air, holding the transparent disk as he proclaimed with conviction, "Behold, the Lamb of God!"

All of us are standing here going along with it, believing in some sense what he's saying, or at least appearing to. And in a moment, we're all going to line up and put a piece of it in our mouths. How odd! I ruminated, as I too slid into the line and filed up to get my tasteless, airy wafer.

These thoughts were fleeting, barely formed, internal ramblings. But they stayed with me as I left the church.

How odd indeed. Bread and wine; food and drink. Both are extremely simple, elemental things, the most basic necessities for survival. And both are *fermented*; both bread and wine undergo that strange process that transforms them from one thing into another. They start out as simple ingredients, insignificant on their own—flour, water, fruit juice—and end up as something else. And it is yeast, a sort of beneficial bacteria, that infects and affects them, blows new life into them like some magical holy ghost, and changes them into something radically different, something better than before.

Food and drink *are* basic necessities of survival but also, if we are fortunate enough, springboards for pleasure that go far above and beyond mere necessity to become a celebration of life itself.

Fine. But body and blood? How does this mediocre wine and flavorless bread become for some people the body and blood of Jesus Christ, someone who lived two thousand years ago? And not just a *symbol* of body and blood but, for true believers, the *actual thing* itself, transformed and transubstantiated through the consecration of the Mass. Are we in the congregation not just allowing ourselves to be pleasantly deceived by these words? How does anything we experience take on more than its superficial impressions and appearances? And, more important perhaps, how can we trust or believe our perception of what we do experience?

The answer to these complex questions is as simple as it is elusive: faith. Your belief makes it happen, and it happens because you believe it. Faith, like yeast, acts as a transforming agent, turning one thing into something else and infusing it with life. Without it, abstract things are like unassembled pieces of a puzzle, unused ingredients, mere potentialities, empty and basically worthless. So how do you believe in something—*anything*; a taste, a feeling, a spiritual concept? That, like the process of fermentation, remains a mystery; you can't necessarily make it happen, or control it or fully explain it. But you can help it along. You *can* create fertile, well-aerated ground in which this transformation might take place, just as a farmer weeds and breaks up the earth around his vines. You do whatever you can and then, as Bartolo said, let it happen and hope for the best. Not everything is totally within our control, and that is as it should be. But if something does sprout up, we should nurture it, enjoy it, cherish it, and experience it as completely as possible.

I did not have any sort of religious revelation that morning in church, but these ruminations did seem relevant to my own present situation. On some level, I thought I could leave my "mess" behind in New York, lose myself by living in a different country and culture, by sweating and freezing in the vineyards, by being with another person. But of course, that was not the case; I brought everything right along with me—*everything*. Rather than losing myself in the vines as I might have hoped to, I found myself there, baggage in hand.

I am lucky. If I am caught in a trap, it is one of my own making and I can get out of it, make it disappear simply by willing it to; by truly believing in myself, in the beauty of life and the world around me. That is easy to say but not always easy to do. I have set out on a long journey, strayed willingly from the straight and narrow path. I still may not know exactly where I'm going or what I'm supposed to do with my life. But at least I know where to look.

That afternoon, the sun was shining, the *nebbia* had finally lifted, and the snow-capped Alps were once again visible off in the distance. There was a certain balminess in the air, one that was familiar but had been absent for a long time. It seemed like the thaw might really stick this time. So I put on my tall rubber boots and went for a walk.

As I headed up the familiar road toward Monforte, the ephemeral late winter sun set off sparks of light on the wet asphalt. When I reached the plateau at the apex of the Bussia just past Cantina Parusso, I turned left onto the curving path alongside the big white buildings of the Aldo Conterno winery and down into the dark, damp shade of the woods. I continued down the steep, slushy path, grabbing onto trees and bushes to keep from slipping into the mucky mess. Everything here was still covered in snow as if left behind in some prehistoric ice age.

I made it down to the flat bottom of the crevice without falling and tiptoed over a log bridge across a little stream rushing merrily between the edges of ice. And there it was. There, over to the left amid the leafless trees and frozen ground, was what I had failed to see before in the lush, verdant spring: the *funtanin!* A pipe poked out of a moss-covered wall of stone and from it water trickled steadily into an old wooden wine barrel cut in half. Another long pipe resting in a notch at the top of the other side siphoned excess water over to another half-barrel slightly below it and from there the water flowed over the rim and down into the slanted forest.

There was a hush broken only by the sound of trickling water. It seemed as if someone or something had been there drinking a moment ago and had run away just as I approached; even the leaves appeared to be crinkling back in the aftermath of the imaginary footsteps.

I went up to the fountain. A plastic cup had been left in a little niche in the stone; I took it, rinsed it, and then filled it directly from the crystal clear stream flowing out of the stone wall. The water was icy cold but light and pure and clean. The flavor was sweet, almost fruity, with an aftertaste of earth and minerals; it was slightly metallic (no sulfur) with a hint of tree-bark resin and green leaves. I had another cup, and then another, standing in the shadowed hush of the crevice.

I had one more taste—it was remarkably refreshing—then replaced the plastic cup in the niche. I felt an odd sense of victory: I had found the *funtanin* and could come back now whenever I wanted. I could even bring empty plastic bottles to fill up and lug home with me.

I peed in the woods as if claiming the spot (they say this water, aside from tasting good, is an excellent purgative) and then made my way up the other side, up out of the darkness into the late afternoon sunlight and into the damp, muddy vineyards.

Unlike the uniformly dark, snowy, wooded pit, here there was an odd dichotomy; in some areas the snow was thoroughly melted while in others it lay as heavily intact as a down comforter. In the places where the snow had melted, there were already soft carpets of pale green grass, above which the wet vines, most of which had already been bent and tied, glistened in the sunlight. These are the good spots, the really good spots, where farmers have learned by long experience to plant their very best grapes (which, around here, means nebbiolo for Barolo). The others, planted to dolcetto, barbera, or chardonnay, still lay in snowy shadow.

Due to the natural contours of the land, these snowless places are not only lit up by the sun but also transformed by it. The sun beating down on these same exact locations for millennia has changed the earth—it is warmer, richer. Organic matter is broken down more quickly and more thoroughly; insects come to bask in its subterraneous warmth, die, and decompose there, adding their mineral richness to the soil.

There is a strange harmony in this apparently free-form chiaroscuro. It seems that the sunny patches and the shady patches (created by nature)—along with their *Vitis vinifera* residents segregated into a strict social hierarchy (introduced by man)—have come to a mutually satisfactory understanding. They all fit together like the pieces of a puzzle and all seem perfectly happy just the way things are. You can feel it: there is a good vibe here, something (as Fabrizio once said about Le Munie) almost sacred.

Looking around at these rolling, vine-covered hills with their nooks and crannies and numerous microclimates, it seems much more complex, much more subtle and personal, than the superficial, quantifiable data alone would suggest. It would appear that grapes, guided by the gentle, intelligent hand of man, express in some special, ineffable but extremely accurate way the particular place and time they grew in, much as people reflect and are a product of where they came from and the experiences they have had.

The good vines (that is, those in the very best positions), basking in the sun on this lovely afternoon at the very first rustlings of spring, appear to be smiling; they seem almost *happy*. Can one say that about a plant? Do vines have feelings or moods? Do they have good and bad seasons just as we have good and bad days?

It seems a silly idea at first. But, feelings or no, vines *do* respond positively to the careful attention of a conscientious farmer and negatively to abuse (hail, deprivation of sunlight, rough handling). They (like people) do best when subjected to a certain amount of stress: they like poor, gravelly soil, uncomfortably steep slopes, and a bit of a breeze; they don't mind struggling for water or having the sun beat down on them day after day. It makes them stronger, better, more interesting. If they have it too easy, if they are grown in flat, rich, well-irrigated soil, they produce flat, boring wines.

It is generally accepted that plants like to be talked to. Many believe that they like music too (if so, I hope they are fond of Bach and Chopin). And vines clearly have an afterlife. As the grape is progeny of the vine, wine is the legacy of the grape, and people have long ascribed human characteristics to it—masculine, feminine, brooding, sexy, petulant, etc. Maybe it will turn out that they were not totally wrong to use such apparently silly words to describe a beverage after all.

Who knows? Evidence of a large, long-extinct pond was recently found on the barren planet Mars. Scientists are all abuzz with excitement at the possibility that other forms of intelligent life may actually exist somewhere in the universe. Perhaps we will find out one day that, in fact, they do exist, that they always have, all around us, right here on Earth. We just never noticed or understood.

By mid-April, the last traces of the snow have melted, leaving the earth damp and fertile (with that, I finally pack the fiber-optic tree back into its box and stick it under the stairs). By the end of May the tears on the branches have dried up and the buds have burst, sending delicate shoots up into the air and turning everything a pastel green. The long annual pilgrimage from vine to grape and, from there, through harvest and fermentation to wine is on again, and it's happening all over the place . . . almost.

Once the earth has thawed, I return with Fabrizio to the empty patch of Le Munie. This vineyard is *not* smiling—it has nothing to smile with; it just sits there helpless and empty, staring off blankly toward Barolo, waiting. But it doesn't have to wait much longer. We get to work.

Manure has been spread out, and the ground turned over to break it up and mix it in. We mark out rows across the length of the vineyard, leaving a wide aisle down the middle. The rows are evenly spaced two and a half meters apart—plenty of room for a tractor to maneuver. Along each row we measure eighty centimeters and mark each spot with a stick.

With the help of Franco and his crew, we dig holes at each marker about a foot and a half deep and a foot in diameter. The ground underneath is dark and cool and moist. This takes several weeks.

In early June the nebbiolo rootstock, all three thousand of them, arrive. They are selected clones from an approved viticultural nursery in Alba, each numbered and certified. These are vines with a pedigree and they are all standing up tall and proud and ready in their plastic trays.

We work our way down the rows on our hands and knees. One by one, pick a plant ("Take the best, the nicest ones, *le più belle*," says Fabrizio. "We have extra here."), loosen it from the tray, place it in the hole, and fill the hole up halfway with earth, making sure the branch is straight and that clumps of soil or stones don't damage it. We then fill the hole with water we carry in buckets from the tractor—there is no longer a well in the vineyard—sloshing all the way. After the water is absorbed, we go back to replace the rest of the soil, leaving a slight indentation around the skinny branch to trap the moisture when it rains.

By the end of June it is done; Le Munie has grapevines in it once again. It will take at least three years for these infant vines to grow big enough to bear real fruit and closer to five before they will reach any significant level of productivity. During this time, they must be nursed and coddled; the soil must be hoed and weeded, copper sulfate *acqua* must be applied. Any plants that don't take in the first few weeks must be replaced. For the next three years, basically everything we did last year with the mature vines—pruning, tying, cleaning, harvest—must be done with these immature plants, even though they will yield no usable fruit. If anything, the infant vines must be coddled even more. They are especially vulnerable right now; they would never have survived a summer like this past one.

Will these vines make it? Maybe yes, maybe no; it all depends. And even if they do, by the time the vineyard becomes productive again, say in four years (plus three more for the time necessary to make Barolo out of these noble nebbiolo grapes), what will the wine industry be like? Will there be a market for Piemontese wine again? Will the grapes be worth anything? Hell, forget about the wine industry—what will the *world* be like in seven years? Will there still *be* one? Or will the divinely delicate balance of the Earth have been destroyed, thrown out of whack by another even more catastrophic terrorist act than the one that brought down the Twin Towers? (Or a nuclear accident, or a widening hole in the ozone layer, or . . . ?)

The answers to these big questions are as unknown and unanswerable as my own little, nagging ones: Where will I be in four years? Will I still be here in Italy for that first real harvest of this new vineyard? What will I be doing with my life then? Will I still be with Ivana? Will I be happy? Questions concerning the future always are unanswerable. You might think you've got it figured out and then in a flash everything changes: *il Destino* strikes, a quick-moving *temporale* blows in from nowhere and wipes out all your hard work—shit happens. Ultimately, it's out of our control, so there are no mistakes to be made, not really, except perhaps not to play your hand at all, to be miserable unnecessarily and not enjoy each and every moment to its fullest while you are in it.

No one knows what will happen, or when or how. But one thing is certain: Fabrizio's digging these holes, planting these new vines, and bringing this old vineyard back to life is not just the resumption of an interrupted family tradition and an engagement with the earth, the culture of a specific place and the natural cycles of the seasons, but also

an investment in the future, in the very *idea* of a future. It is a coura-
geous act of faith, just like the one that mysteriously transforms the
crappy wine into the Cup of Salvation and the insipid wafer into the
Bread of Heaven, just like the one that induces me to whisper *"Ti amo"*
into Ivana's ear when she wakes up in the morning and allows her,
turning over sleepily and embracing me, to reply, *"Ti amo anch'io."*

Finally, we finish filling in the holes. With that, the baby vines stand
as straight up from the curvature of Le Munie as the spines on a sea
urchin, all perky and eager, like young kids in formation ready to
march into school for the first time, full of anticipation, a little afraid
perhaps, but ready to confront whatever might be in store for them.
They stare wide-eyed up at the new sky and introduce themselves to
the sun: *Here we are!* they seem to say, as they slowly start to sink their
nascent roots into the dense, unfamiliar soil and shudder imperceptibly
as the cool, refreshing air drifts up the valley and swirls around them
with a gentle caress. A whole new cycle has begun in this old vineyard.

"Bon," Fabrizio says once we pack the dirt around the last two
vines. *"Anche questo è fatto."*

We stand up, dust the earth off our knees, and make our way up
the steep path to the top. When I get there, I turn and pause for a moment to survey the new vineyard. Looking down on our work, the
words of the teary-eyed old farmer among his vines come back to me:
"I put everything I had into this place, my whole life. Now I have nothing; I'm only half a man." And my response: "No, you're a whole man.
You made this place and it is beautiful, *un bel pezzo*. You should be
proud. Everything you put in here is still here, in the earth, and always
will be no matter what."

I understand now how he felt. Whatever happens in the future, the
work I did in this vineyard and the time I spent here will always remain
in some way, in the earth and in the vines, in the grapes and in the wine
that is made from them, just as this place and the experiences I had
here will always remain within me.

"Cosa dici, andiamo?" asks Fabrizio, patting me on the back.

"Sì," I respond. *"Andiamo."*

I go home, wash up, change my clothes, and then head back over to the Mascarellos' house beneath the vine-covered ridge of the Bussia. When I get there, the long table is already set, bottles of wine are open, and everyone is gathered: Fabrizio, Marilena, and baby Pietro, Sebastiano and his wife, Teresina, and Tiziana. Argo is spinning circles in excitement and anticipation and Andrew the cat watches calmly from atop a ledge. Even Bruna and Fiore stop by with a truffle he and Mickey stumbled upon *fuori stagione*. Everyone is smiling and talking and laughing; spring is here, the new vineyard is planted, life is good. Spirits are high, as well they should be: a celebration is about to begin—a celebration of the *living*—and everybody's invited.

Suddenly, a car comes rolling up on the gravel. Ivana jumps out, slams the door, and rushes in, smiling and breathless. "I'm starving," she says. "Let's eat!"

NOTES

Prelude

1. The communes of Castiglione, along with Serralunga and Barolo, are completely within the boundaries of the Barolo wine zone; only parts of the eight remaining communes—Roddi, Grinzane Cavour, Verduno, Cherasco, La Morra, Novello, Monforte, and Diano d'Alba—lie within the approved area. Only wines made from the nebbiolo grape produced within the delineated zone and meeting other criteria for yield and aging may be labeled "Barolo."

4. Francesco

1. The Consorzio di Tutela Barolo, Barbaresco, Alba, Langhe, e Roero is the organization officially entrusted to administer the DOC regulations (see note 3, chapter 30, pages 326–27). *Tutela* means safeguard or protection, and the Consorzio di Tutela is a nonprofit organization comprising primarily the producers themselves. Originally founded in 1934, long before the DOC/DOCG regulations themselves were formulated, it is one of the oldest such organizations in Italy.

5. Vines and Wines

1. Some believe that the Etruscans, in what is now the region of Tuscany, already had viticulture when the Romans got there. That may well be true, though, if it is, the Etruscans likely learned how to grow grapes and make wine from the same source the Romans did; that is, from Greek traders or perhaps directly from the region in the Middle East where grapes are thought to have had their origin.

2. In the region of Campania in southern Italy there is an ancient wine, originally produced by Greek settlers, whose DOC discipline still requires that

the vines be planted in the traditional manner next to poplar trees. As the vines grow, they crawl up the trees; harvest takes place on tall ladders at heights of up to forty feet!

3. Grapes *will* naturally ferment in this way; however, most modern wine-makers prefer to add specially chosen cultivated yeasts to better control fermentation. For more on fermentation, see page 123 and pages 209–13.

4. For example, commercial viticulture and winemaking were first brought to California largely by Italian immigrants in the late nineteenth century.

6. Into the Vineyard

1. Most vines in the Barolo zone are pruned using what is known as the Guyot system.

7. Blendings

1. The tradition of *mezzadria*, a kind of sharecropping system in which people lived on and worked (but did not own) the land in return for giving half the bounty to the *padrone*, was not officially abolished until 1963.

2. Castiglione Falletto was the site of the area's first and oldest *Cantina Co-operativa*, located on an estate called Montanello, which is now a winery and *agriturismo*. Here farmer-members would bring their grapes for vinification. At the base of the Castiglione hill is a co-op called Terre di Barolo, which is one of the most active cooperative cantinas, with hundreds of small farmer-members. During harvesttime, there is a long, long line of tractors bringing grapes to the cantina.

9. Tongues

1. Grapes not suitable for Barolo can, however, be used for other, "lesser" denominations such as Nebbiolo d'Alba or Langhe Rosso.

2. Generally the adjacent communes of La Morra and Barolo, with their sandy, chalky soil, produce more elegant wines with a touch of finesse and slightly higher acid. The wines of Castiglione, Serralunga, and Monforte, on the other side of a valley where the soil tends to be more compact and clayey, are more muscular, dense, rotund, and powerful. The other six communes align themselves with one or the other of these two stylistic factions. However, these are big generalizations.

3. Santo Stefano Perno is one of Monforte's top Barolo crus. Giuseppe Mascarello, Bruno Giacosa, and Valentino Migliorini (Rocche di Manzoni) are among those who have parcels here.

4. This crevice runs below the long narrow strip of a vineyard called Rocche (which literally means the "fortesses" but here indicates rocks or cliff),

which runs from Castiglione into Monforte. The vines tilt steeply down to the east, then stop; the trees at the edge quickly give way to a sharp cliff and a deep drop down to near-impenetrable forest.

10. Tying the Knot

1. For the most part, vines can be trained to bend wherever you want, though haphazard direction is often an indication of inferior pruning.

11. Rites of Spring

1. *"Trovare l'America"* is a phrase used throughout Italy. Most of the Italians who left during the big wave of immigration around the turn of the twentieth century, however, were from southern Italy where conditions were especially desperate. The Piemontese are often referred to, even by themselves, as *bugia nan* (nonmovers): even in the worst of times, life is pretty tolerable in Piemonte and that—plus their inherent hard-headedness—tends to make them stay put. Notable exceptions to this are the well-known author from Asti, Cesare Pavese, who wrote about his travels to America (mostly *after* he returned to Italy, however) and a bunch of Piemontese who emigrated to Argentina. They remained; and to this day there are communities of people in Argentina who, along with their Spanish, speak a perfectly preserved Piemontese dialect!

13. Better Late Than Never

1. Technically, I wasn't applying to *obtain* Italian citizenship, but rather submitting a request to be officially recognized *(riconosciuto)* as a citizen. According to the law, I already was an Italian citizen by blood right; I just needed to obtain the necessary documents and go through the proper bureaucratic processes to prove it.

15. Renza's

1. Almost everyone here wears slippers (called *pantofole*) at home or, in some cases, at work. There is an expression in Piemontese, *"Sté an pantofole,"* which means to make oneself comfortable or at home. *Un pantofolaio* can also be used derogatively to refer to someone who is indolent or lazy.

2. Others say the house is called this because the person who lived there was named LaMorte.

17. Bearings

1. The "re-" refers to the re-creation of Italy as a single political entity on the peninsula, such as had existed prior to the fall of the Roman empire.

2. The best answer I got was offered by Castiglione's most notable celebrity resident, the singer-songwriter-raconteur (and stationmaster of the Cuneo train station) Gianmaria Testa. We were out on Renza's terrace drinking wine one afternoon. Because many of his lyrics are in Piemontese, the subject of language naturally came up, and I posed my question.

"Yes, it is a bit peculiar, but I think I have an answer," he said. "Do you know the church bells that ring before the main Mass on Sundays? First, at ten o'clock a little bell rings after the hours to remind everyone it is Sunday and the Mass is going to take place in an hour. Then they ring again at ten thirty and once again at ten forty-five. Then, just before eleven, three big bells sound—*bong! bong! bong!*—like a final reminder that you'd better get yourself up to the church or else! Now, you must remember that in those days, Mass was not just a religious obligation but also an occasion for the whole village to meet, a chance for everyone to see one another and exchange news. Even men (like my father) who didn't actually go inside for Mass would go up and congregate outside the church. Because of the importance of Sunday Mass as a major social event, people became accustomed to speaking of the three bells that immediately proceeded it—*tre bot: padre, figlio, e spirito sancto*—and naturally transferred that to the daily ringing of the campanile at one, two, and three o'clock. That's why they would never think to say *quart bot*—it simply doesn't exist."

3. Oddly enough, local pizzerias, the majority of which are owned and operated by transplanted southern Italians, also specialize in the preparation of fish dishes, though most of it is frozen.

4. *Razza Piemontese* meat has what is known as a *Denominazione di Origine Protetta*, or DOP (protected name of origin), which is very similar to the *Denominazione di Origine Controllata* used for wine.

5. Trichinosis is almost nonexistent in Italy.

6. Antipasti are small appetizer-like dishes that precede the *primo* or pasta course(s).

7. Peppers from Carmagnola are especially renowned; they have their own DOP and their own autumn festival. They totally dominate local markets during the autumn harvest season, like an invasion of bright red and yellow balloons just barely contained in their plastic net-covered crates.

8. In the "olden days" prior to the post–World War II affluence, people practically lived off polenta. To this day, other Italians sometimes refer derogatively to the Piemontesi as *polentoni* (big polenta eaters).

9. The hectare (*ettaro* in Italian) is the standard unit of agricultural measurement in countries using the metric system. One hectare equals 10,000 square meters, or 2.47 acres.

10. A *brenta* held 50 liters, and its conformity was strictly regulated by law.

11. Nebbiolo is grown in other areas of Piemonte, where it is known as Ghemme or Carema or Spanna, as well as in the Valtellina area of Lombardia, where it is made into five crus: Sassella, Vagella, Grumello, Inferno, and Sforzato.

18. A Tale of Two Cities

1. Giulia's mother was related to Louis XVI of France, and her father was a descendant of Jean-Baptiste Colbert (1619–1683), one of Louis XIV's ministers.

2. Camillo Ludovico Borghese (1775–1832), of Rome's aristocratic Borghese family, was a general in Napoleon's army. In 1803 he married Napoleon's sister, Pauline. From 1807 through 1814 he lived in Torino and served as governor-general of Piemonte, which was then under French rule. Upon the fall of Napoleon, he separated from his wife and went to Florence, where he died in 1832.

3. Palazzo Barolo, Giulia's former residence, is home to the Opera Pia foundation, which is open (though sporadically) to visitors.

4. Giulia Falletti di Barolo wrote a number of books in French (later translated into Italian) on her childhood memories, travels in Italy, and experiences with prison work.

5. The former royal palace at Verduno is today a hotel, restaurant, and cantina. In 1849, Carlo Alberto, king of Piemonte and Sardinia, abdicated, and the throne passed to his son, Vittorio Emanuele II, who would later, with the help of Cavour, become the first king of the unified Italy.

Vittorio Emanuele II had also taken a liking to Barolo, for in 1858 he purchased property in the commune of Serralunga d'Alba, where he created an estate both to house his mistress, Rosa Vercellana (the so-called Bella Rosin), and to make wine. Called Fontanafredda, the estate still produces wine (it is currently owned by the Monti di Paschi di Siena bank), and it is possible to tour the still-intact former royal cellars.

6. Camillo Benso di Cavour (1818–1861), born into an aristocratic family in Torino, was named after Camillo Borghese, who, along with his wife, Pauline, sister of Napoleon Bonaparte, acted as Cavour's godparents. He served as a page at the court of Carlo Alberto and from there went on to become principal architect of the unification of Italy and the first prime minister of its constitutional monarchy. Throughout much of his adult life, Cavour administered the family estate in the town of Grinzane, where he served as mayor from 1832 to 1848. The castle is now home to Enoteca Regionale, Piemonte's first regional center for wine tasting and promotion.

7. Giulia and her husband, Carlo, have been officially beatified, the second step on the road to sainthood. Their neighbor and fellow winemaker, Cavour, did not fare so well; he was excommunicated by the Vatican for his key role in the reunification of Italy, which severely curtailed the church's political influence.

8. There is another man who played an important role in Giulia's life. In 1830 Silvio Pellico, an Italian author and political activist, was released from prison in Spielberg, Austria, where he had been incarcerated for agitating against the Hapsburgs, and shortly thereafter wrote a memoir of his experi-

ences called *Le mie prigioni* (*My Prisons*, various editions). Giulia, with her abiding interest in the plight of inmates and commitment to prison reform, read his book and initiated a meeting.

The Falletti later offered Pellico a position as librarian in their Barolo castle (it was probably Giulia's idea). After her husband's death in 1838, Giulia never remarried but Silvio stayed on, and the two lived side by side in the big castle for the next sixteen years, right up until his death in 1854. Many naturally assumed the two were having an amorous relationship. This is certainly possible, even probable, but they never overstepped the bounds of propriety; whatever occurred between them was private and personal and remains so to this day.

19. *King of the King of Wines*

1. Bartolo Mascarello may be a traditionalist when it comes to wine making but he does not live in the past. He is a political activist, a longtime socialist, anti-Fascist, and former member of the Resistance; he is as much against certain political figures as he is against the use of small French oak barrels in the making of Barolo, and doesn't hesitate to voice his opinion on either subject. In 2001, Bartolo designed a whimsical and controversial label for his wine with the words "No Barrique/No Berlusconi!" and a photo of the politician. When the label first appeared, the bottles were sequestered by the police because of the image of the prime minister. Bartolo responded by adding a flap with a ladybug on it covering up Berlusconi's face. This was enough to placate the authorities, but the flap is easily lifted to reveal the face underneath, as if hiding behind an innocent insect.

Bartolo Mascarello died on March 12, 2005. The funeral took place two days later just outside the gates of Barolo cemetery. Nearly all the prominent Barolo producers were there along with numerous farmers, neighbors, representatives of the wine *consorzio*, and aged colleagues of the Resistance, their banners lifting in the gentle breeze. Though, in accord with Bartolo's wishes, a church Mass did not precede the outdoor ceremony, a portable organ played and the parish priest offered a blessing.

It was later learned that the Mascarello home was burglarized during the funeral while the whole village was busy paying their respects at the ceremony.

20. *Vanishing Species*

1. The Hudson Valley and Finger Lakes regions of New York also have a long history of wine making, but never quite bounced back after the repeal of Prohibition.

2. With the shrinking demand for wine, it seems as if hazelnuts might be making a comeback. New trees are being planted in many of the larger flatter

tracts of land at the base of vine-covered slopes, tracts that might have been designated for grapes or poplar trees. Hazelnuts are the second most lucrative crop after grapes and are sought after by the local confection industry.

21. War and Peace

1. In Barolo, the irrigation of vineyards is prohibited as it mitigates the climatic effects of a given vintage.

2. Small amounts of sulfur are actually present in the protective yeast layer that naturally forms on the skin of wine grapes.

22. Inferno

1. In the end, a shocking total of 15,000 people died in France during the heat wave of 2003.

2. Some people install fine-mesh nets that can be unfurled over the rows in the event of a storm, but their use is not widespread and their effectiveness is questionable.

23. Party On!

1. Camiot's house has been painstakingly renovated and is now a small elegant hotel; the owners, however, were thoughtful enough to retain the cutout corner of the door where his dogs once roamed freely in and out.

24. Green Harvest

1. One quintale equals 100 kilograms, 1 kilogram equals 2.24 pounds, and 1 hectare equals 2.47 acres.

26. Harvest

1. *Rasp' di San Martin* ("Saint Martin's clusters) are small secondary grape flowerings on the upper part of the vine that don't ripen until much later (if they ripen at all). Around Saint Martin's Day, November 11, thrifty farmers used to go through the vineyards again to collect these late grapes and make highly acidic wine for their own consumption.

27. Tasting Terroir

1. From the 2001 vintage on, Parusso's three Bussia vineyards will be blended into a single *cru* called simply Bussia.

2. Each commune in the Barolo zone is thought to have its own unique characteristics, as are certain subzones within the communes. This is true of

other wine regions as well, such as the individual villages of Chianti Classico or of Burgundy.

3. Whether to bottle individual *crus* or blend them into one wine depends on many things, such as a given winemaker's vineyard holdings, the vineyards themselves, the vintage, wine-making philosophy, logistics (if a cantina has several small plots and large fermenting tanks, they may have to blend in order to fill a tank—needless to say, it is an additional difficulty and expense to keep numerous lots separate), and marketing. Some time ago, *crus* were popular and winemakers were encouraged to make a lot of them. Today, due to a crowded marketplace and the confusion generated by multiple bottlings from the same producer, *crus* are falling somewhat out of fashion.

28. Realm of the Senses

1. If wine is too cold, as many white wines often are, its flavors and fragrances will be muted. (Take it out of the ice bucket, place it on the table, and watch it blossom.) Red wines should not be served too warm for the same reason. Young, aggressive, acidic red wines often do well with a slight chill.

Many experts suggest that big tannic wines, such as Barolo or Bordeaux or Amarone, be opened several hours in advance in order to let them breathe. I, however, generally prefer to open such bottles just before dinner, perhaps decant, and then watch them unfold and develop throughout the meal.

30. Into the Cantina

1. Parusso does something a bit unusual; after harvest they often leave the grapes in a big ventilated, temperature-controlled room for a day or so to let them "relax," cooling them down to about 12 degrees Celsius (just over 50 degrees Fahrenheit). They believe this permits a better separation of stem from grape during the subsequent destemming process, thus minimizing harsh tannins and making for a better extraction of the juice. Being able to lower the temperature of the grapes before vinifying them is especially advantageous in an extremely hot year like 2003, and ventilation can also help dry them out in the event of rain or dew.

2. Because nebbiolo is a late-ripening grape and the weather just after harvest is often quite cool, one of the common problems with the wine prior to the technical innovations of Marchesa Giulia and Louis Oudart was excessive sweetness and volatility due to prolonged and incomplete fermentation.

3. *Denominazione di Origine Controllata* is the Italian equivalent of the French AOC *(Appellation d'Origine Contrôlée)* system. It is the official codification of the particular qualities of a particular type of wine from a specific place. The criteria of "typicity" are based on local wine-making tradition and defined in large part by the producers themselves. They include as many of the mitigating factors of wine production as possible, including grape variety (or varieties), delimitations

of the zone, yield per hectare, aging, vinification, and labeling requirements, minimum alcohol content, and in some cases even trellising. As with any official system of rules and regulations, the DOC laws are often criticized by producers for being outdated and overly restrictive. They do change, albeit slowly and usually after the fact, to reflect changing ideas about wine and wine making.

Italy's DOC system is extremely complicated. There are a staggering number of denominations, and the list is always growing. To have a wine recognized as a DOC, producers must make their case to the national Italian Ministry of Agriculture and then be extremely patient; it can take years to be accepted or refused.

Barolo belongs to a small group of elite wines known as DOCG. The G refers to *"garantita"* (guaranteed), which is a superior category with even more stringent regulations.

As if the world of DOC was not already complicated enough, several years ago a whole new category called IGT (*Indicazione Geografica Tipica,* or indication of geographical typicity) was established to regulate many new grape varieties and wine-making practices in a somewhat looser, more all-embracing structure.

4. Following the second or malolactic fermentation, dead yeast cells and other deposits sink to the bottom of the barrel and form a thick sediment called lees. Usually this sediment too is racked off by siphoning the wine from one barrel to another. Some producers, however, prefer to leave it in the cask during the aging period in the belief that this gives the wine a certain creaminess and enhanced complexity. The French term for aging on the lees is *sur lie.* If the sediment is not racked off, it needs to be stirred up periodically with a big rod to mix it with the wine. This is called *bâtonnage.*

5. Many producers are less scrupulous, opting to bottle the inferior wine— which does, however, meet the minimal legal requirements—as Barolo or, worse, sell it off *sfuso* (in bulk) for some commercial producer to do the same. This enables them to cut losses and bring inferior Barolo to market at a lower cost, but wreaks havoc with consumer expectation of both quality and price and does inestimable damage to the reputation of the appellation.

31. Subterranean Treasures

1. There are ongoing efforts to try to track the genetic DNA of the Alba white truffle so as to differentiate it from other types, create a DOP (*Denominazione di Origine Protetta,* or protected place name of origin), and protect it from confusion with "impostors." But they have not been successful thus far.

33. Trials and Tribulations

1. Of course, in the United States we have similar rules pertaining to foreign visitors, which, after 9/11, are being more stringently enforced than ever. As an American on my home turf, however, I never had to deal with them.

35. Cycles Old and New

1. Most who employ *barriques* mix new and used barrels for their Barolo. After that, they use them once more for nebbiolo or Barbera and then discard or sell them.

2. Well, almost no one. In the town of Verduno is a small historical cantina where they have never completely stopped. The grapes from their tiny piece of the Monviglero vineyard are crushed by foot in the old-fashioned method. Why? Because that's how they've always done it. I myself have reserved a spot at next year's harvest.

36. Local Landmarks

1. "The Lord of hosts has a day against all that is proud and lofty, against all that is lifted up and high; against all the cedars of Lebanon, lofty and lifted up . . ." (Isaiah 2:12–17). This tree, however, does not strike one as being offensive to the Lord of Hosts, and I suspect it would be spared if and when such a day should come.

2. In fact, as I discovered when I was invited to a wine tasting and finally got inside it, the cube is hollow; you walk in through glass doors and then down a flight of stairs into the subterraneous cantina. The wines, however, are quite good.

39. Day of the Fat Bulls

1. Carrù is solidly outside the Barolo zone. A Barolo or Barbaresco could certainly be drunk with *bollito*, but it might well overshadow the food. My favorite match is a Dolcetto di Dogliani such as the one from the nearby Cantina Clavasana. Its simple, robust fruitiness pairs well with the simple, rich, full-flavored meat, and the nice firm acidity cuts right through the fat. The wine's simple structure and soft supple tannins provide a lovely supporting-actor role to *bollito*, letting the meat/condiment flavors come shining through.

Dolcettos from the Dogliani appellation generally have a leaner structure with more rusticity, acidity, and tannins than Dolcetto d'Alba, which tends to be rounder, fruitier, and more voluptuous. One of my favorite producers of Dolcetto di Dogliani is Chionetti, whose wines are unusually complex and elegant.

41. Infusions

1. Examples include Genepy, made in the mountains near the French border from wildflowers called gentian, and vermouth, an infusion of tree bark and other secret ingredients, which was invented in Torino. Martini & Rossi, one of the first commercial producers of vermouth, invented the martini in the 1800s to promote their product.

2. A fortified wine is a wine to which alcohol, often brandy, is added. This addition both "fortifies," or strengthens, the wine by raising the level of alcohol and stabilizes it. Port is a well-known example of a fortified wine.

3. Quinine is derived from the cinchona tree, indigenous to the Andes Mountains in South America. It was first "discovered" in the early seventeenth century by an Augustinian monk who noticed that native Indians ground the bark of the tree to use as a remedy for fever and other ailments. Afterward, it became quite popular throughout Europe both as a medicine and, with its refreshingly bitter, herbaceous flavor, as a component in various beverages and tonics.

acino. An individual grape berry.

acqua. Water. In the context of viticulture, it means water mixed with chemical products (*prodotti*) such as copper sulfate, used to help prevent disease in the growing vines.

agriturismo. Owner-operated tourist-based business in a rural area. These are typically informal places to eat and/or sleep where visitors can enjoy regional cuisine and get a more intimate, first-hand experience of the surrounding area than otherwise possible.

albeisa. Name of the distinctive type of bottle developed during the 1700s and traditionally used for wines from the area around the city of Alba. The shape resembles that of a Burgundy bottle, with thick, dark blue-brown glass, a punt in the bottom, and the word "ALBEISA" embossed across the shoulder.

ampelography. The science of identifying, mapping, and describing the various varieties of *Vitis vinifera*.

annata. Vintage. A given viticultural year, from pruning to harvest, with its particular combination of meteorological factors and, by extension, the wines produced from grapes of that year.

azienda. A commercial enterprise. *Azienda agricola* is an agricultural enterprise (such as grape growing). *Azienda vinicola* is a winery. *Azienda vitivinicola* is a business that both grows grapes and makes wine.

Barolo. One of the eleven *comunes*, or villages, of the Barolo wine zone, it gave its name to the finest wine made from the nebbiolo grape.

barrique. A small (225-liter) barrel, usually made of oak. Barrels of this kind are traditional in France and are now widely used throughout Europe and America.

bâtonnage. A French term indicating the technique of stirring up the lees in a wine barrel by using a metal stick *(bâton)*.

botte. A large (5,000-liter) wine cask traditionally used in many areas of Italy, often made of Slavonian oak or chestnut. (In Piemontese, a large wooden cask is called a *carera.*)

brenta (Piemontese). A long cylindrical container made of wood that was used to transport wine.

Brix scale. A scale that measures the percentage of sugar in grape juice, named for Adolf Brix, a nineteenth-century Austrian scientist.

buta giù (Piemontese). Literally "throw down." This term is often used in place of the Italian *diradamento* to refer to the practice of removing some grape clusters from each vine prior to harvest and throwing them away, thereby reducing the yield and, it is believed, improving the quality.

cantina. A place where wine is made; winery.

capannone. A large shed or warehouse used for commercial purposes.

capo a frutto. A fruiting cane. The branch that, during pruning, is selected to propagate buds for a given growing season.

capo a legno (*portao* in Piemontese). A spur with two or three buds that is left on the vine during pruning to generate a fruiting cane for the next season.

ceppo. The thick lower stalk of a vine; that is, the part below the branches.

ciabòt (Piemontese). A shed or small house in a vineyard, used for storage of small tools and equipment and/or temporary shelter.

cin cin. A toast, often made when clinking glasses. "Cheers!"

comune. Municipality.

concime. Fertilizer. *Concimare* means "to fertilize."

contadino. A farmer.

cru. A French term (literally "growth" or "crop") widely adopted throughout the wine world to signify wines made from grapes of a single, specific vineyard.

damigiana. A demijohn. A large glass bottle with a 55-liter capacity, used to store wine. Traditionally, the bottle was wrapped in straw.

diradamento (Italian) or *buta giù* (Piemontese). The practice of reducing a vine's yield by discarding a number of grape clusters prior to harvest.

DOC/DOCG. Wine appellation. The initials stand for *Denominazione di Origine Controllata* (Controlled Place Name of Origin) or the more prestigious *Denominazione di Origine Controllata e Garantita* (Controlled and Guaranteed Place Name of Origin). A set of official regulations that seeks to define a particular type of wine and assure that any wine called by that name meets certain minimal standards and criteria.

enoteca. The place where wine is sold; wineshop. Many *enoteche* also offer the possibility of wine tasting (*degustazione*).

fermentazione. Fermentation. The chemical process by which sugar is converted to alcohol.

filari (Italian) or *filagn* (Piemontese). Rows (of vines). Often used to refer to a vineyard.

fili di ferro. Metal wires strung the length of a row to support the vines.

gemme. Buds. Nodules along the vine that open during the growing season to produce additional branches, flowers, and leaves. The first opening of the buds after pruning in a given growing season is referred to as "bud break."

gich (Piemontese). The tip of a vine branch.

giornata piemontese (Italian) or *giurná* (Piemontese). A traditional agricultural unit of measure equal to 3,810 square meters.

girapoggio. Literally, "go around the hill." The method of planting vines horizontally along hillsides.

grappolo. A grape cluster.

gurat (Piemontese). Supple tips of willow branches cut from the tree and used to secure vine branches to wires and poles during *tòrse*. Nowadays *legacci* (ties) of paper-covered wire or string are usually used.

Guyot system. The method of pruning most often used in the Barolo area, named for Jules Guyot, a nineteenth-century French scientist. Other methods of pruning include the cordon spur and the pergola.

IGT (Indicazione Geografica Tipica). Indication of Geographic Tipic-ity. A lesser appellation than DOC because it is more general and less stringent.

innesto. Graft. The technique of splicing one type of vine (known as the scion) onto the rootstock of another.

legare. To tie. (See *tòrse* below.)

lievito. Yeast. A bacterium that, whether it occurs naturally or is artifi-cially selected and induced, effects the process of fermentation.

macerazione. Maceration. The practice of letting the grape skins re-main in contact with the juice following destemming and crushing but before the actual process of fermentation has begun.

manoal (Piemontese). A manual, often unskilled, laborer.

mezzadria. A sharecropping system in which peasants lived and worked on land owned by someone else in exchange for a share of the food they produced there.

mosto. Must. Grape juice, with or without skins, prior to fermentation.

nebbiolo. The most important grape variety of the Langhe area.

pali. Poles (made of wood, cement, or metal) arranged in rows along which wires are strung to support grapevines. Originally, *pali* were made of wood branches; newer ones are made of cement or metal.

piegare. To bend, as in bending a vine branch prior to securing it to the wire. (See *tòrse* below.)

pigiare. To remove the stems from the grape bunches after harvest prior to fermentation. *Pigiatrice* (destemmer) is the name of the modern piece of machinery that removes the stems from grape clusters and, in the process, breaks the berries.

pintun (Piemontese). A large bottle of wine, such as a double magnum or 3-liter container.

pompare. To pump grape juice through various stages of the vinifica-tion process via rubber tubes and a hydraulic pump. The pump is also often used to perform racking.

potare (Italian) or *puè* (Piemontese). To prune or cut back the previous year's growth from a vine. This is the beginning of the annual cycle culminating in the autumn harvest of grapes.

produttore. Wine producer.

pulidé (Piemontese). Cleaning or removal of extraneous branches dur-ing late spring.

quintale. One hundred kilos (about 220 pounds).

rasp' di San Martin (Piemontese). Small, immature grape clusters on the upper part of the vine that are not picked during harvest, but are collected much later on, traditionally on Saint Martin's Day (November 11). This was also the day when vineyard workers and grape farmers were traditionally paid.

ritocchino. Vine rows that are planted straight up and down a hill (as opposed to horizontally across it; see *girapoggio*).

scalusé (Piemontese). General spring cleaning and repair of poles and wires in a vineyard.

scarzolé (Piemontese). Practice, sometimes known as "green harvest," of removing extraneous green leaves and small branches from the vines in late spring or early summer as the grapes are beginning to mature.

sfuso. Loose or bulk. A term referring to wine that is sold in bulk (truck, barrel, demijohn), as opposed to wine sold in bottles.

spumante. Foamy. A term that refers to a sparkling (as opposed to a still) wine.

sur lie. A French term that denotes aging wine on the lees (yeast deposits) instead of racking them off.

tappo. A bottle stopper traditionally made of cork (*sughero*). *Cavatappi* is a corkscrew, while the expression *"Sa di tappo"* refers to a wine that is corked.

terroir. A French term now used internationally to refer to the unique combination of factors—climate, soil, exposure to sunlight and air, etc.—that play a significant role in a given vineyard.

tirare su i tralci. To lift up the vines.

torchio. A machine for pressing the juice out of the grapes. Often bunches of white grapes are placed directly in the press, and the juice is fermented later. Red grapes that are macerated and fermented on the skins are run through the press later to squeeze out the remaining juice.

tòrse or *tòrze* (Piemontese). Similar to the Italian word *torcere* (to bend or twist), this is the second critical step in the Guyot system of pruning and trellising, in which the fruiting branch (*capo a frutto*) is bent (*piegato*) horizontally and tied (*legato*) to a wire.

tralcio. A branch of a vine.

travaso. Racking. Moving wine from one vat to another just after (or at the tail end of) fermentation to aerate and remove sediment. Racking may also take place later on during the aging process.

uva/uve. Grape/grapes.

uvaggio. Blend of different grape varieties in a given wine.

vendemmiare. To harvest. *Raccogliere l'uve*, or to collect grapes, is also used. *Vendemmia verde* (green harvest, a k a *diradamento* or *buta giù*) refers to the practice of cutting off some grape clusters prior to harvest to reduce the yield and improve quality.

vigna. A small vineyard or patch of vines. *Vigna vecchia*, a designation sometimes seen on wine labels, refers to old vines.

vigneto. Vineyard.

vinacce. Pomace. The leftover skins and seeds after pressing; they are often used to make grappa.

vino. Wine. The alcoholic beverage resulting from fermented grape juice.

vite. Vine. The crawling plant, *Vitis vinifera*, that has been cultivated for millennia to produce grapes for wine. (Not to be confused with *vita*, life or waist.)

vitigno. A particular variety of grapevine such as nebbiolo, barbera, dolcetto, or cabernet.

Vitis vinifera. Family of so-called "noble" European grapevines.

LIST OF RECIPES

INDEX

Entries in small capitals indicate recipes.